LITTLE
MAN

"At once a ruthlessly honest personal story and a devastating critique of contemporary American culture ... What makes [Tizon's] writing compelling is his ability to investigate and explain complex topics, deftly weaving in information from websites, history texts, university research, and social media, combined with intense self-examination. His willingness to look inward gives him more authority to unpack some of the damaging misperceptions about Asian men."   — *Seattle Times*

"Highly readable ... This personal narrative of self-education and growth will engage any reader captivated by the sources of American, and Asian-American, manhood — its multitude of inheritances and prospects."   — *Minneapolis Star Tribune*

"Part candid memoir, part incisive cultural study, *Big Little Man* addresses — and explodes — the stereotypes of Asian manhood. Alex Tizon writes with acumen and courage, and the result is a book at once illuminating and, yes, liberating."
                            — Peter Ho Davies, author of *The Fortunes*

"In *Big Little Man* Alex Tizon fearlessly penetrates the core of not just what it means to be male and Asian in America, but what it means to be human anywhere."
                            — Cheryl Strayed, author of *Wild*

"An unflinchingly honest . . . beautifully written, often discomforting examination of Tizon's remarkable, yet thoroughly relatable, life." —*New York Times*

"A well-paced, engaging combo of history, memoir, and social analysis . . . Tizon's skill as a feature reporter serves the book well, producing a narrative that moves fluidly between subjects, settings, and gazes." —*Publishers Weekly*

"A deft, illuminating memoir and cultural history." —*Kirkus Reviews*

"Written compellingly . . . eye-opening . . . deeply felt, extensively researched." —*Booklist*

"Tizon's candid journey into the shifting and multiplying definitions of manliness and the masculine ideal is revelatory and sobering." —*Library Journal*

BIG
LITTLE
MAN

BIG
LITTLE
MAN

In Search of My Asian Self

Alex Tizon

Mariner Books
Houghton Mifflin Harcourt
BOSTON • NEW YORK

First Mariner Books edition 2018
Copyright © 2014 by Alex Tizon

For information about permission to reproduce selections from this book,
write to trade.permissions@hmhco.com or to Permissions,
Houghton Mifflin Harcourt Publishing Company,
3 Park Avenue, 19th Floor, New York, New York 10016.

hmhco.com

Library of Congress Cataloging-in-Publication Data
Tizon, Alex.
Big little man : in search of my Asian self / Alex Tizon.
pages cm   Includes bibliographical references.
ISBN 978-0-547-45048-3 (hardback) | ISBN 978-0-544-23285-3 (ebook)
ISBN 978-1-328-46014-1 (pbk.)
1. Tizon, Alex. 2. Filipino Americans — Biography. 3. Young men — United States —
Biography. 4. Asian American men — Social conditions. 5. Asian American men —
Psychology. 6. Asian Americans — Ethnic identity. 7. Sex role — United States.
8. Masculinity — United States. 9. Sexual attraction — Social aspects — United States.
10. United States — Ethnic relations. I. Title.
E184.F4T59 2014   305.895'073 — dc23   2013045020

Book design by Chrissy Kurpeski

Printed in the United States of America
DOC 10 9 8 7 6 5 4 3 2 1

For Melissa, Dylan, and Maya

Contents

BIG
LITTLE
MAN

1

Killing Magellan

What is the knocking at the door in the night?

—*D. H. Lawrence*

When I was twenty-nine, I flew to the island of Cebu in the Philippines to watch a fight. I arrived on a sweltering morning with nothing but books and some clothes in an overnight bag, which I threw into the trunk of the first taxi that stopped for me. It was a white clunker with the words LOVE DOLL COACH painted in red cursive on the passenger door just above a phone number and a smaller inscription that read RIDE NICE IN PARADISE. The driver's name was Bobby. For the next two mornings, Bobby greeted me with "Morning, Sir Alex."

"No need to call me 'sir,'" I'd say.

"Yes, sir, Sir Alex," he'd assure me.

He was not being funny. Bobby went about his work with robotic swiftness and a prefab smile that appeared on cue. He was eager but detached at the same time, with no interest in making real contact. I never felt at ease around him. I wanted to be friends; he wanted to be my servant. I would learn that his was the prescribed demeanor for all service workers in the Philippines. Saying "sir" two or three times in a single sentence was not considered excessive. Chronic obsequiousness had seeped into the national character during four centuries of colonial rule. Bobby also chain-smoked, which no doubt contributed to turning the whites of his eyes crayon red. He had greasy misshapen hair and grimy fingernails as long as guitar picks. He was not a pretty sight in the morning. But I knew no one else, and there were so many other things to rest my eyes upon.

Cebu is one of the Philippines' larger islands, a long skinny outcrop of sand and forest in the central region known as the Visayas. From the air, the island resembles the profile of a high diver in mid-dive: from the tip of the toes in the north to the fingertips in the south is 120 miles. The thickest part, around the trunk, is twenty-five miles across. If the plunging diver were facing east, the capital — Cebu City — would be right near the belly button, which was where Bobby drove me on that first day. It was an hour's drive in heavy traffic from the airport.

To get to my hotel, the Love Doll Coach wound through a labyrinthine maze of crowded neighborhoods, the *masa* made up of brown bodies in scant attire pressed closely together and yet somehow also moving like a river. Up we chugged on climbing narrow streets lined with rickety storefronts, their corrugated steel awnings bent and rusting out. Freshly butchered goats hung from hooks, blood still dripping from open mouths. Women in shorts and flip-flops with baskets of fruit on their heads walked past with small children running in orbits around sinewy legs. I rolled down my window and was instantly smothered in air thick with exhaust and something else, what was it—sweat? The smell of toil. Occasionally an ocean breeze cut through, and a hint of wet sand and palm trees. The scent of mangoes from somewhere.

It was all new to my Americanized senses. I was awash in new-ness, as if I had landed on a never-discovered continent. And yet it was not my first time here. I was born on one of these islands. My blood, with its tinctures of Malay and Spanish and Chinese, came from the same pool as those of the masses we passed on the road. At age four I was brought by my parents to America, a land where people did not look too kindly on a groveler, for instance, anybody who said "sir" three times in a single sentence. I recognized Bobby because I had a little bit of Bobby inside me, and I didn't like it much. Becoming an American meant I had to hate the groveler and exorcise him from my soul. It was hard work, becoming an American, and I felt I'd succeeded for the most part.

Yet I was not "all-American." I could never be that. Most of us, when imagining an all-American, wouldn't picture a man who looked like me. Not even I would. You would have to take my word for it that more than a few times in my life I looked in a mirror and was startled by the person looking back. I could go a long time feeling blithely at home, until a single glance at my reflection would be like a slap on the back of the head. *Hey! You are not of this land.* Certainly during my growing-up years in America, many people, friends and strangers, intentionally and not, helped to embed in me like a hidden razor blade the aware-ness of being an outsider.

I remember an encounter with a fellow student at JHS 79 in the Bronx, where my family lived in the 1970s. I was about thirteen. My school was just off the Grand Concourse on 181st, a five-story brick building with bars over all the windows and dark clanging stairwells that might as well have been back alleys. Some stairwells you did not dare travel alone, but I was new and didn't know better. One afternoon in one of these stairwells, an open hand with five impossibly long fingers fell hard against my chest and stopped me in my tracks.

"What you supposed to be, motherfukka?" the owner of the hand said.

"Wha—?" I stammered.

"Are you deaf, boy? I said, What you supposed to be?" The owner of the hand was a tall black guy, Joe Webb, who turned out to be the oldest and biggest member of my seventh-grade class, a man among boys. He was one of those guys whose muscles bulged like rocks sewn under skin, whose glare conveyed the promise of apocalyptic violence. "Are you a *Chink*, a *Mehikan?* What?"

Blacks, Puerto Ricans, and other Latinos made up the majority of students at the school. There were some whites and a handful of Chinese and Taiwanese. I was the only Filipino in the school, and a lot of students like Joe had never met one and knew nothing about the Philippines. I told him what I thought I was.

"You don't look American, bitch," he said. He eventually let me pass after I gave him the change from my pocket, which I would learn was really what he was after. Moving around the school meant paying certain tolls. Sweet scary Joe Webb. We ended up sitting next to each other in English, and he would copy off my test answers with my implicit consent. After he had found me acceptable six months into the school year, he became my friend and protector for the rest of my time at JHS 79. Sometime later that year, when another kid tried to shake me down in

the very same stairwell, Joe loomed over him with those murderous eyes and long fingers rolled up into fists and the other kid
melted into the darkness whence he came, never to bother me
again.

Joe's original query was a question I've been asked in various,
usually more tactful ways ever since I could remember. *What
you supposed to be?* From where on this planet did you come?
What *are* you? The person in the mirror was the color of coffee with two tablespoons of cream. The face was wide with hair
so black it sometimes appeared blue. The eyes were brown and
oval, the nose broad, the lips full. A face that would have blended
naturally with those I saw on the street that morning from the
back seat of the Love Doll Coach.

On my second day in Cebu, Bobby arrived at my hotel at the
predetermined time, cigarette dangling from his lips. "Sir, good
morning, sir." The cigarette bobbed.

He knew where I wanted to go, so with as few words as possible we took off, winding back down the hill and through the
center of town. It was a hot, cloudless morning, and the shops
and food stalls were just beginning to open up. The Love Doll
Coach wove through traffic, passing colorful jeepneys with their
passengers peering out from glassless windows and buses that
left huge black plumes of smoke in their wake. Motorcycles and
tricycles darted in and out of small spaces between vehicles.
What people called tricycles were actually motorcycles with
covered sidecars, essentially mini-taxis that could carry one to
twenty people, depending on how tightly the passengers were
willing to fold their bodies. One tricycle we passed carried what
looked like an entire class of young girls, their plaid uniforms
flapping in the wind, their arms and legs intertwined and cling-

ing to various parts of the sidecar and one another: five in the cab, four on the roof, three behind the driver, and another three riding the rear bumper. One of the girls on the bumper mugged for me as I snapped a picture. She must have wondered what was so interesting.

I took pictures obsessively. It was my first time in Asia since my family had left twenty-five years earlier. Taking photographs at least gave me the sensation of absorbing some of what I was seeing, gave some reassurance that the images would be filed in a safe place so I could retrieve them and make sense of them later when I wasn't so wound up. Maybe all those images in aggregate could help me remember what I had forgotten. The camera became an extension of my body, which had become a sensory organ taking in everything—too much to process, too much to respond to in the back of that chugging taxicab. For long periods I became mute, maxed out to blankness.

The Love Doll Coach came upon a narrow steel bridge arcing over a wide expanse of slate-gray water. On the other side, like a piece of driftwood, was Mactan Island. We crossed the bridge and took a road cutting through the island's center, passing villages that teemed with street vendors hawking everything from frog-skin purses and coconut-shell brassieres to ukuleles handcrafted from fine mahogany, and then we headed in a straight line toward the swamps of the eastern coastline. Up ahead in the middle of the road, an old man walked bent over with a dozen guitars strapped to his back, the fangled load held together by a filigree of blue nylon. Bobby flicked his wrist and went around him at sixty-five miles an hour, muttering something as he passed, his tone suddenly not so servile, the cigarette bobbing.

The car slowed to a stop along a lonely stretch of scrub. Bobby lit another cigarette. I looked at him blankly. He blew a gust of smoke, smiled a fake smile, and pursed his lips together into a point so that lips protruded farther than nose, and I realized he was pointing at the Magellan monument, which appeared like a lone gray sentinel in the distance. "There, Sir Alex," he said simply. I got out to look.

The air smelled of low tide and charcoal; somewhere meat sizzled over a fire. The sound of faraway cars dissolved into the tidal swishing that passed for silence on these islands. This was what I had come to see. The structure, behind a wrought iron gate, stood forty feet high, made of stone and divided into three tiers, the uppermost a long, slender spire, like the top of an ancient church. Weeds sprouted from the craggy seams. I stood at the edge of the gate, gazing up at the spire and then looking around at the vacant beach, vaguely disappointed.

It was on this spot that a battle took place almost five hundred years ago, one that for half my life had occupied a dimly lit corner of my imagination. It was a fight I had long wanted to visualize up close. On one side of the battle were bearded alabaster-skinned men in armor with swords and spears made of iron. On the other side were men who looked more like me: compact, sinewy, ebony-haired, copper-skinned men who wore hardly anything at all and whose weapons were made of bamboo and stone. At the time I could not articulate exactly why it was important to stand in that spot, to feel the same sand beneath my feet, other than because the victors of that battle five hundred years ago were the men who looked like me.

I had come to believe that men like me did not win contests with other men. Men like me—yellow- and bronze-skinned sons of Asia—could not stand up to the power of Western men, those white-skinned gods who marched through history conquering everyone in their path. Men of Asia were small and weak and easily vanquished. They lost wars. They let their nations be conquered and dominated. They allowed their women to be turned into groupies and whores. The men of Asia had no choice. They went limp in the face of Western power.

And they brought their weakness with them across the oceans to the new lands where they settled. In the America that I grew up in, men of Asia placed last in the hierarchy of manhood. They were invisible in the high-testosterone arenas of politics and big business and sports. On television and in the movies, they were worse than invisible: they were embarrassing. *We* were embarrassing. The Asian male in cinema was synonymous with nebbish. They made great extras. In crowd scenes that required running away, Asian men excelled. They certainly did not play strong male lead roles, because apparently there were no strong Asian males with sex appeal. On the public sex appeal scale, Asian men did not even register. They were hairless, passionless, dickless. Tiny minions. Houseboys.

All these strands made up the mythology, one that was all the more potent because it was mostly unspoken. It did not need to be spoken. But I didn't see it as a mythology for a long time, most of my growing up. I experienced it as a set of suspicions that seemed corroborated by everyday life. I could not point the finger of blame at anyone. Mine was an education that came from the air itself. At school, it was as much what was *not* taught. Asians simply did not come up in history class, except as victims who needed saving (Filipinos, South Koreans, South Vietnamese) or as wily enemies who inevitably lost (Chinese, Filipinos, Japanese) or as enemies who managed not to lose by withstanding mind-boggling casualties (North Koreans, North Vietnamese). Asia was a stage on which dynamic Westerners played out their own dramas and fantasies, with Asians as incidentals. I graduated from high school unable to name a single preeminent East Asian figure in history who was a force for good.

Admittedly my education was spotty, downright poor in some grades, and this was due in part to my family's vagabond way of life. But everywhere we traveled, I met better-educated students who knew even less than I did about East Asia and its descendants. And everywhere I saw Asian men in domestic work and manual labor, in jobs that no one else wanted to do. They were in restaurant kitchens wearing hairnets, washing dishes, busing trays, and making sure the trash never overflowed. They were apple pickers and ditch diggers, the ant-like workers in backroom sweatshops and basements, the slaughterhouse grunts in kill chambers, the raincoated slimers scraping entrails into noisy grinders. They changed the sheets in hotels and hospitals, swept the streets and raked leaves for people who had more important things to do. Most Asians I encountered in our cross-country travels were gardeners, seamstresses, laundry operators, or janitors, people who cleaned up other people's dirt and kept their heads bowed in obeisance and in one way or another said "sir" for a living.

So ingrained was this mythology of the Asian male that when, as a teenager, I heard the words uttered by the character Song Liling in *M. Butterfly,* "I am Oriental. And being Oriental, I could never be completely a man," I could only shut my eyes in recognition and shame.

Yet I also knew it wasn't true. Somewhere in the middle of myself, like a corpuscle hidden under layers, I knew the mythology was a sham. I had lived too many secret moments in which I felt iron within me, and I recognized the same thing in my father and brothers. These were glimmerings. They beckoned me. Looking back now, I see that I needed more evidence to bolster the case, to feed the secret hope.

At age fourteen I began keeping files. Figurative files in my head, but also actual file folders with headings such as "Great Orientals" and "Asians in the News" and "Oriental vs. Asian?" scribbled in big Sharpie letters. Whenever I ran across anything fileable related to Asians, in particular pertaining to race and manhood and power and sex, I would make a note and tuck it away in one of the folders. I would clip articles from newspapers and magazines, make copies of reports, tear out pages of books. The folders became fat and unwieldy. They sprouted sub-folders. The collection outgrew its ratty cardboard box and ended up in two metal file cabinets that my mother and I picked up at a garage sale. One was beige, the other black, and they sat side by side in our garage like two miniature office buildings.

"What's in those?" one of my little sisters asked.

"Files," I said cryptically.

"Of . . . ?"

"Top secret information," I said. It was hard to explain.

In truth, I could not explain it for a long time. The files were like evidence folders for a deeply personal, almost spiritual investigation, one that had no working title. It was like the globs of mud and weed that hornets gathered to build their nests around our house, something they did by instinct. I gathered information out of a sense of urgency that I could not identify. Years would pass before I realized that all the gathering was an effort to ease my inconsolable secret. I wasn't building a house or nest. I was trying to construct a key, something to unlock the door into belonging. Now I know there was no such single key but all manner of keyholes I could peer into for clues. One keyhole appeared in the tenth grade, in Mr. Thompson's social studies class.

Mr. Thompson was the school's band teacher, but in the tiny eastern Oregon school I attended for two years, teachers pulled double and triple duty, often teaching subjects they knew nothing about. A short, roundish man with a ruddy face and bright blue eyes, Mr. Thompson ran his social studies class the way we students believed we liked it: he left us alone and said hardly a word during the entire period. We pulled colored information cards out of a box, read a card, and then answered a short list of questions—all on our own. Students generally saw the class as an opportunity to doodle or write notes or nod off.

One afternoon, during a section titled "The Great Explorers," I pulled a card on Ferdinand Magellan, the Portuguese explorer who, under the Spanish flag, attempted the first circumnavigation of the world. The card informed me that the "Captain General," as he was called, made it halfway. During a stop in the islands that would eventually be named *Las Islas Filipinas* in the region now known as Southeast Asia, Magellan was murdered in a skirmish with island natives. On a piece of scratch paper I wrote *"Magellan in the Philippines—skirmish"* next to a stick figure with two *X*'s for eyes. Later that day I tucked it in the file folder titled "?" and promptly forgot about it.

Four years passed before I saw that piece of paper again. I was a freshman in college living in Eugene, Oregon, an eclectic community of mill workers and artists and student counterculture types to whom I found myself drawn for a while. At Christmas break I drove home to see my family and to stock up on any loose items I could pilfer from my mother's garage (my parents had divorced by then). And there, behind some orange barber chairs stacked like the Leaning Tower of Pisa, were my file cabinets, just where I had left them. I opened the drawers and glanced at the tab titles, and then got the idea that I could sure use some extra file folders at school, so I grabbed a few and began riffling through them to see what I could throw out. I pitched most of what was in the "?" folder, but the piece of paper with the stick figure, for some reason, I folded up and stuck in my wallet. And there it remained for a few more months, until one night among the stacks at the University of Oregon's Knight Library, in a bleary-eyed fit of procrastination, I pulled it out and stared at the scribbled words.

On the study table were several pages of Algebra 101 equations that awaited my attention. In my hand was the scrap of paper. The scrap of paper won. I spent the rest of the evening tracking down books on Magellan and then skipping to the parts that described his demise. It turned out that the circumstances of his death amounted to more than a skirmish. The fifteen hundred native warriors involved in the fight were led by a chieftain named Lapu Lapu, about whom almost nothing had been written. I got an incomplete in Algebra that quarter, but a glimmer came to me that night in my lonesome corner at the library: I conceived the notion that I needed to travel to Mactan.

Another decade went by before the opportunity arose. One late afternoon, while I was working as a reporter for the *Seattle Times*, a Filipino American teacher I knew, Gloria Adams from Meany Middle School, called me on the phone. Tita Glo reminded me of my mother, and from the first moment we met she treated me as if I had spent eighteen years in her house and had just left home yesterday.

"You always tell me how much you want to go. Now we're going and you're coming with us," she said resolutely.

"It's sort of short notice, Tita Glo," I said. (*Tita* means "aunt" in Tagalog.)

"Tell them you're doing a story. . ."

"About what?" I said.

"What do you mean 'about what'? About a group of Filipino American teachers going back to their homeland to network with Filipino teachers, like a cultural exchange kind of thing. How does that sound? It's good! It's human interest! You can write a positive story for once. You're coming."

Against my better judgment I broached the idea to my editor, who didn't wait to hear the end of my pitch before she said, "I think you should go." The *Times,* my editor said, was trying to get more people of color, besides criminals and athletes, into its pages.

Three weeks later I was on a Korean Air flight with Tita Glo and her colleagues, crossing the Pacific Ocean to a country I knew only in ethereal fragments, through bits of conversation and snapshot images and letters. I was going ostensibly to write a feature story for a newspaper, but I had my own separate and secret mission. Crammed in my suitcase were four books on Magellan's epic voyage, and folded into one of them was a yellowed scrap of paper with a picture of a dead stick figure. After two weeks with the teachers, I split off on my own and flew to Cebu, where upon landing I immediately met the man who would become my unofficial tour guide.

Now, I walked back to the Love Doll Coach and found Bobby asleep, a cigarette butt hanging precariously from his lips. I would let him sleep a little longer. In the back seat was a pack containing the Magellan books. I reached in through an open window, grabbed the pack, and headed back toward the monument. There I found a bench next to a small palm tree whose branches gave me slices of shade, and one by one I read the accounts of the battle.

The sun, pale and fierce, had risen to the middle of the sky by then. The shouts of children floated in the air. On the beach a Filipino couple strolled barefoot, holding hands. She wore a long wispy white dress that blew about as she walked. His hair was almost as long as hers, and from one of his fingers dangled their sandals. She whispered something to him and he bumped her away affectionately, still clinging to her hand. She bounced back into his arms, her expression expectant, and with her slender neck leaning into him, she whispered something else. Beyond them the sea faded into sunlight.

The invaders must have appeared like visions in a dream, a nightmare—human forms covered in metallic skin, long red beards hanging from gaunt faces, and their eyes opaque with menace. They carried shields and swords and lances unlike anything ever seen in these lands. The attackers numbered only fifty, but they marched toward the beach with an attitude of invincibility. At the head of the contingent was a man shorter and darker than the rest, one who walked with a limp, and yet he moved forward with the calmest visage, lips curled into a sneer. The islanders must have known instinctively he was the one called *Magellanes*.

"Watch how the Spanish lions fight," Magellan had said before disembarking from the longboat and beginning the march to shore. Watch and learn. See how we butcher these *indios*. Hear their cries for mercy. We will show none.

Not all of Magellan's men shared his confidence; some thought the attack unnecessary and foolish. The island's topography was unknown to them. Anything could happen. But then again, many in Europe thought the whole idea of circling the globe foolhardy, and Magellan only seemed energized by the skepticism. He had banked his reputation and fortune on this voyage. His Armada de Molucca had crossed two oceans and withstood famine and disease and violent clashes with native tribes in search of the oceanic route to the Spice Islands, a mythical place that, with its trove of spices under the crown's control, would make Spain the wealthiest nation in the world, and Magellan rich and famous beyond mortal dreams. Eighteen months after leaving Seville, the armada's men had made landfall in these islands, subduing with their thunderous cannons all the tribes they encountered. It was part of their routine: to see the natives quake at the sound of their blasts and then preach the gospel to them.

All acquiesced, except one. Lapu Lapu had sent a message to Magellan that his people—the inhabitants of Mactan—would never submit. Magellan vowed holy war, and just to show his seriousness, he sent an advance team to burn down one of Lapu Lapu's villages. An untold number of Mactan inhabitants were killed by the blaze. Then at dawn on April 27, 1521, Magellan and sixty of his men and hundreds of native allies sailed toward Mactan to finish the job. As they approached, Magellan sent an emissary to shore with a message for Lapu Lapu: "If you would simply obey the king of Spain, recognize the Christian god as your sovereign, and pay us our tribute, I would be your friend. But if you wish otherwise, you should wait to see how our lances wound."

Lapu Lapu sent the emissary back with his own message: "I submit to no king and do not pay tribute to any power. Our warriors have lances, too, and they are made from stout bamboo and hardened with fire. Come across whenever you like."

Magellan attacked. His soldiers paddled toward Mactan but

discovered that the island's coastal shelf was unusually shallow, forcing them to anchor more than a half mile from shore. Magellan ordered his native allies to stay in their boats and watch how the Spanish lions fought, and then he picked forty-nine of his men, in full armor, and began wading to shore, their distant glimmering silhouettes presenting a sight never seen before by the Mactanese waiting on the beach. The attackers must have experienced their own alarm when they saw the size of Lapu Lapu's force.

A compact, thinly muscular man with jet black hair to his shoulders and tattoos of suns and triangle patterns over half his body, Lapu Lapu, at about age thirty-five, was already a hardened warrior. He and the men of his tribe had been battling Bornean marauders and Moluccan pirates since they all were old enough to hurl a spear, and Lapu Lapu proved adept at making war. Throughout the Visayas he was known as fearsome and unpredictable. As Magellan and his men waded toward shore, Lapu Lapu's warriors shouted war cries, jabbed their spears skyward, and rushed headlong into the shallow waters to meet the invaders.

This surprised the Europeans. They had thought the battle would begin on land, where armor would be an advantage; in the water, their metal plating weighed them down and slowed their movements. The two armies clashed in the shallows and bodies fell, turning the waters an inky red. For more than an hour the battle line moved back and forth, but the Mactan warriors' numbers and ferocity proved overpowering.

"They shot so many arrows at us and hurled so many bamboo spears and stones that we could scarcely defend ourselves," wrote Antonio Pigafetta, the armada's chronicler, who was among the attackers.

The Europeans were pushed into deeper water. Magellan ordered a retreat and gallantly stayed and fought as his men slogged back to the boats. With the Spaniards' numbers dwindling, the few left behind were overwhelmed. Magellan suffered several telling wounds in rapid sequence: a poison arrow struck his right leg, a sword slashed his left leg, a rock violently knocked off his helmet, and then, the coup de grâce, a bamboo spear pierced his face. The Spaniards watched in horror as their Captain General fell before the warriors. Pigafetta: "They immediately rushed upon him with iron and bamboo spears and with their cutlasses, until they killed our mirror, our light, our comfort, and our true guide."

The surviving Spaniards retreated to Cebu, where they were met with more violence. In an earlier stopover, the Europeans had helped themselves too greedily to the native women, had shoved their weight around too freely among the men. The Cebuanos feared the Europeans' weapons. But word of Magellan's defeat at the hands of Lapu Lapu had traveled quickly. The Europeans had been exposed as mortals. Emboldened Cebuano warriors lured the European survivors to a feast and then ambushed them, killing twenty-seven. The rest of the crew barely escaped to their ship. They continued westward, and eighteen months later completed the first recorded circumnavigation, although at a high cost. The Armada de Molucca left Spain with five ships and 237 men; three years later, one listing ship with eighteen emaciated and disease-ridden survivors made it home, none of them able to say precisely what had happened to the remnants of their Captain General's body.

I threw the books into the back seat, startling Bobby, who had reclined his seat and was deep in slumber. He shot upright and began fumbling with the keys even before he realized I was already seated behind him. I tapped him on the shoulder and gave him another start.

"Yes, Sir Alex," he blurted. "Go where now, sir? How about the mall, sir?"

"How about the hotel. I'm done," I said.

"Are you sure, sir? You still have one hour, sir," he said, referring to our agreement to a three-hour tour.

"No, I'm tired. Magellan wore me out," I said.

"Yes, sir, Sir Alex," he said.

"I'm not going to miss you 'sirring' me, Bobby."

"Yes, sir," he said, backing up the Love Doll Coach and nearly mowing down a group of young men passing by. The group parted effortlessly and moved along the contours of the car like water around a rock. Fascinating, I thought. Such different rules, such a strange world. Back home, Bobby's action would have elicited a curse and a middle finger, if not a pounding on the hood. These boys didn't bother to glance at Bobby. As far as I could tell, the only rule of the road that everybody recognized was that the bigger object had the right of way, pedestrians and stray dogs be damned.

I flew out of Cebu the next day as gray clouds began moving across the island. After a brief layover in Manila, where I picked up the rest of my luggage, I boarded another plane for the marathon flight over the Pacific and back home to Seattle. It was only after we ascended into the spotless blue space above the clouds that I relaxed into my seat and let out a long, slow breath. I pulled out a thick stack of just-developed photographs and began lazily flipping through them. There was so much to think about.

I had thought my visit to Mactan would be the highlight.

After I'd reread the accounts of the battle and imagined how it must have looked on the beach that morning a half millennium ago, I walked around the Magellan monument, studying it from different angles. It was erected just up the beach from the spot where the Captain General had fallen, and I tried to imagine the swarm of hands that ripped at his armor and clothing, the cries of victory from fifteen hundred mouths.

Standing at the entrance gate, I waited for crowds to come, school buses to empty in the parking lot, the uniformed children swarming around the monument that doubled as a gravestone for one of the most famous explorers who ever lived, but they never came. Someone told me later that large crowds rarely gathered there. It had to be the world's loneliest tomb of a great explorer, and a part of me was tempted to think the sonovabitch had it all coming: his gruesome death, the ultimate humiliation among the natives he wanted to subjugate, his subsequent dishonor in Spain, his memorial on this desolate stretch of mud and mangrove.

See how the Spanish lions fight, my ass. This is what I wanted to say, was prepared to say, but gazing at the sad gray stone sapped the venom right out of me. I waited for a feeling of exultation — an Asian warrior had prevailed here, after all — some solitary inner version of the raucous celebration in black America when Jack Johnson defeated Jim Jeffries for the heavyweight championship of the world in their historic 1910 match in the Nevada desert.

But Lapu Lapu's triumph did not bring the inner celebration or solace I had imagined. There might have been a whisper of validation somewhere in my core, but another part of me felt unfastened and vaguely put off by the idea, which I must have been carrying around unconsciously, that stabbing someone in the face and cutting up his body amounted to some kind of ultimate proof of manhood. At the very least, whatever gloating I could muster felt temporary and fragile, like a house of cards on a windy beach. Maybe all gloating is that delicate. Something or someone always comes along and knocks it all down.

In any event, I came away from my Mactan adventure with a sense that my floating notions of what made men *men* needed other anchors. This was a dawning that seems to occur earlier in many young men. I became aware of the possibility that I had succumbed to an insidious kind of racial envy, that my secret quest was simply an attempt to assuage a feeling of inferiority that was mine alone. Maybe it was not an Asian manhood thing; maybe it was just a *me* thing.

I had begun the investigation long before I realized it *was* an investigation. It was never very organized or even clear-hearted in its motivations. It felt as much a groping in the dark as a quest to create something new. I was trying to piece together a three-dimensional puzzle without a picture of what it should look like. But eventually a picture did begin to emerge, and I met other men and women — in books, chat rooms, coffee shops; in conferences and lecture halls; in pool halls and back alleys and bars — who'd been working on their own puzzles and who would help me place a stray piece here and there. What I found at the end of this quest was both astonishingly hopeful and stubbornly uncertain. Because, of course, what emerged was a moving picture, changing year by year, changing even as I speak.

I flipped to the picture of the couple on the beach. They seemed so detached from the rest of the world, so caught up in their own moment. I wondered whether they ever entertained such conundrums, whether he ever questioned his manhood, or she questioned his questioning. *You just grew up on the wrong continent!* a college friend once told me in one of the few drunken conversations that actually had import. We laughed. Maybe he was right. I put the pictures away, shut my eyes as if for the first time in days, and woke up in America.

2

Land of the Giants

They have gone across,
melting away on the other side

—*Li Wei*

When did this shame inside me begin? Looking back now, I could say it began with love. Love of the gifted people and their imagined life; love of America, the sprawling idea of it, with its gilded tentacles reaching across the Pacific Ocean to wrap around the hearts of small brown people living small brown lives. It was a love bordering on worship, fueled by longing, felt most fervently by those like my parents who grew up with America in their dreams. The love almost killed us.

My family arrived stateside in 1964, the beginning of a turbulent time in the land of dreams. A handsome and charismatic president had been killed by an assassin's bullet in Dallas five months earlier—shot in the head with his beautiful wife sitting next to him in a Lincoln Continental convertible. Images of war in a faraway place called Vietnam flickered across the television screen. And soon there would be riots in the cities, demonstrations on college campuses, armies of angry people clashing in the streets over issues we did not understand. Civil rights? What are those? "You people," I remember my father telling the television. "You have everything. What reason do you have to be unhappy?" It was a sentiment he would repeat often during our first years here. He was referring to Americans, in particular white Americans. My parents used "white" and "American" interchangeably. They saw them as the same.

We had nothing. We had crossed the Pacific on a Pan Am jet with all our belongings in cardboard boxes. My parents had borrowed money for the trip. Borrowing became a mode of survival they would never escape. At the time, of course, they saw it as temporary, an unavoidable step. As soon as they got a foothold in this great country, they would pay back their debtors and climb their way to wealth and happiness, the dream to which my parents clung as only émigrés from a poor country could — with everything on the line, no reserves, no plan B. The only plan those first years was to survive.

We landed in a strange and beautiful place called Los Angeles. We bought a car, our first, a slightly banged-up white Plymouth Valiant station wagon. From the back seat I saw wide streets lined with palm trees all so uniformly spaced apart and sidewalks clean and smooth like gray glass, the stucco mansions planted like jewels in the rolling hills, and the people — so many beautiful people, tall and robust with their long pointy noses and bright smiles and big round eyes the color of the afternoon sky. My mother swore she saw Henry Fonda coming out of a Hollywood restaurant.

MY MOTHER: So elegant!

MY FATHER: That wasn't Henry Fonda, Mama.

MY MOTHER: Never mind.

It wasn't, he insisted. They spoke in a hybridized language of Tagalog and English. She claimed to have spotted George Peppard, too. I didn't know who these people were, and in any case, everybody looked glamorous, magnified, like they had just stepped out of a movie screen, ten feet tall and cartoonish in their perfection.

I pressed my face against the glass as I saw black people for the first time, on a playground, their shirtless bodies gliding up and down a basketball court. *"Neh-gros,"* my mother said matter-of-factly. It was the accepted term in the Philippines and was still widely used in the United States at the time. Occasionally I would catch glimpses of people who vaguely resembled us, with their compact bodies and tawny skin and black hair. Mexicans, I would learn. We had much in common with them, *indios* with Spanish blood and culture running through our veins.

We stayed eight months in Los Angeles, living out of our cardboard boxes in a rented bungalow, as my parents scrambled to find a place to settle. One evening my father announced over a bucket of Kentucky Fried Chicken that we were moving north, to a city none of us had heard of. Somewhere near Canada. He had landed a job there. The very next morning, my parents, four kids, and an aunt, Lola, who had crossed the ocean with us, packed ourselves into the Valiant, an open roadmap on the dash. My father gripped the steering wheel, his mind on a thousand possible outcomes, while the rest of us drifted in and out of sleep. I watched drowsily as the valley turned flat and the flatlands turned hilly and the mounds turned ocher and the meadows turned mountainous — Mount Shasta, Mount St. Helens, Mount Rainier, with night intervening sometime between them — and eventually the mountains gave way to lush, dripping green. I shut my eyes near Tacoma and opened them near Seattle. My parents pronounced it *Shia-tel*.

They made airplanes here, my father informed us. The jet that carried us all to America was made somewhere in the middle of this place. By *Boy-ying*, my mother said. I think it's Boe-*wing*, Mama, said my father. *Boe-wing. Shia-tel.* What kinds of words were these? What kind of place was this?

My parents secured a small house, a $90-a-month rental, in the Roosevelt Way area on the city's northern outskirts. It was the only white house on the block and the only white house we would ever live in. It had a sagging roof, walls that leaned, floors and stairs that creaked under layers of linoleum and carpet. We all marveled at the fireplace. None of us had ever seen an indoor fireplace before. It was the deal-sealer for my father. I hold an image of him gazing at it, dreaming up all the scenes of family happiness that would play out in front of a blazing fire. A hearth of our own in America.

It took days of frantic phone-calling for my parents to come up with the deposit and first month's rent. How they would pay for the second month, they didn't know. But they found a way. The second, the third, the fourth. My father worked two jobs. During the days he worked at the Philippine consulate. His official title was assistant commercial attaché, but mostly he played host and tour guide for visiting Filipino dignitaries. His night job was cleaning trailers at a trailer park in Snohomish County. My mother got a part-time job dissecting and analyzing rat brains at a local medical lab. She rode two buses to work. They came home exhausted every night. Their combined incomes never seemed enough.

"The bills keep piling high, and Papa and I don't have fifty dollars to our name," my mother wrote in her journal in the fall of that year. My mother would keep a daily journal for fifty-two years, until she was too feeble to pick up a pen. "Neither of us can sleep at night because we don't know what will happen tomorrow."

Somehow every month the ghost of Saint Rita, patron saint of impossible dreams, to whom my mother prayed, would come to the rescue: a loan would come through, an advance would be approved, a rich relative would take pity. "Another miracle!" my mother would write. During these high times, my mother would come home with a new dress from Sears and my father would arrive with a new pet. At one point we had a white German shepherd, a white rabbit, two parakeets, and two young robins that had fallen out of a tree. Animals and children made my father happy, as long as he didn't have to do the work of taking care of them. And so we made it to Christmas that year, clinging to a wisp.

Two days before Christmas, snow fell. Our Roosevelt Way neighborhood turned into a postcard scene before our eyes, happier, lighter, all the hard edges cushioned, all the dings and divots covered in a dreamy blanket. I remember holding my face up for the longest time, feeling the snow alight on my skin and hair and eyes, and just letting it melt over me. There were sounds I had never heard before: the soft crunch of boots stepping onto a virgin patch of snow, the oceanic swishing of tires, waves upon waves, the harsh scraping of shovel blades against rough concrete. Without asking, one of our neighbors, an old stooped white man with white hair and a red plaid hunting cap, shoveled the sidewalk in front of our house and didn't turn back to acknowledge us when we said thank you.

Ganyan sila, my father would say. "That's just the way they are." Presumptuous but kind. One old woman across the street occasionally brought us flowers from her garden and never hesitated to correct our pronunciation. "It's *Rosa-velt*, dear. Not *Roos-belt*."

The snow brought us into a moment of knowing exactly where we were. Up until then, life in the new land had been a fierce running-around, a blur. But the snow turned everything still, and we knew we were really in America. Even at that young age I knew we were lucky. We were among the divinely selected few who'd slipped through the eye of the needle, and we had arrived intact, frazzled but whole. At least for a few hours it felt that way. It was Christmas Eve and we would celebrate.

"Smells good!" my mother said, coming home from work and catching a whiff of the stuffed chicken Lola had put in the oven. My mother had "done" fifty rats that day. She was tired and glad to be home, giving each of us a hug over our fat winter coats.

"What did you do with them?" my little brother asked.

"*Wh-i-i-i-shhhhht,*" my mother said, her finger drawing a quick straight line across my brother's skull. He chortled and ran off. A short time later she was putting Frank Sinatra on the phonograph when my father came home from the trailer park, looking disheveled but with a broad smile and a bag of groceries. "How are my beautiful children?" he said, sweeping up my two-year-old sister Ling in his arms. Hugs, tousled heads, my brothers Arthur and Albert and I jostling one another, Lola scurrying around the kitchen preparing various dishes, my sister shrieking with joy as our father started a fire in the fireplace. The evening proceeded dreamlike, Frank Sinatra crooning of chestnuts

roasting and sleigh bells ringing, all of us inebriated with gratitude, my father enhancing his gratitude with Budweisers. The
fire kept going out, and he kept feeding it with anything that
looked combustible, and soon the rest of us joined in, collecting cardboard and loose papers and dead branches from outside
and throwing them on the embers. Nothing worked.

Finally, my father went out and returned a short time later
with what looked like a package of giant Vienna sausages. "Pres-
to-Logs!" he announced triumphantly. "America!" He put all
four in the fireplace, lit them, and before long we had a roar-
ing fire. We ate dinner in front of a blaring television, the room
brimming with chatter. We opened presents — my younger
brother and I got toy Winchester rifles, just like Lucas McCain's
in *The Rifleman* — and hooted and hollered and shot each other
to death until two a.m., my brother and I falling asleep in our
closet-turned-fort, still clutching our weapons.

Around four a.m. my father heard Ling crying. He got out of
bed and immediately tipped over face-first onto the floor, struck
timber. He gasped for air, as though a belt had been pulled tight
around his chest. "What's happening!" he said out loud. He
struggled to his feet and called out for my mother. She jumped
out of bed and immediately toppled to the floor. Papa stumbled
to the next room and found Lola and Ling on the floor, gasping
for breath, both drenched in what we would later learn was their
own urine. When he could not rouse them, he ran to wake up
Arthur, who took a couple of steps before his legs gave way and
he slumped against a chair, coughing with his eyes closed. My
father almost did not make it to Albert's and my room, nearly
succumbing in the hallway. He saw a window and kicked it open
and then ran to find my brother and me in the closet, our gap-

ing mouths contracting in grotesque rhythm—"like perch in a bucket," he said later. He opened another window and another until every window in the house was swung wide. He got on the phone, his body now wracked with convulsions. "Something's happened. Please come," he was able to say before vomiting on the floor.

I don't remember anything of the incident or immediate aftermath. My siblings and I were knocked out cold. I started coming out of the fog in the afternoon, but my mother told me I was not completely coherent until the evening of the next day. All of us had terrible headaches. Our house appeared as if an army had marched through it. I found out what had happened mostly from snippets of talk between my parents and the seemingly endless number of people who visited that week. A few had read about it in the *Seattle Times*, which ran a six-paragraph story on the front page of the local section with the headline: ATTACHE HERE AVERTS FAMILY YULE TRAGEDY. My mother clipped the article and later read it aloud to us:

> A cozy Christmas fireplace scene almost ended in tragedy for a Philippine consulate official and his family here. Francisco A. Tizon, assistant commercial attache at the consulate, his wife, their four children and an aunt nearly were overcome by carbon-monoxide fumes from smoldering sawdust logs.

The most interesting line came at the end of the third paragraph, which stated that before going to bed, my father had closed the fireplace draft.

"*Sunog ang ay patay na.* The fire was dead," my father would say. "I looked inside and saw this piece of metal . . . a lever . . . and I pulled it and something closed. I thought, 'Better closed to keep the cold out.' The flames were out anyway. I didn't know.

How could I know?" Indeed, how could a man who'd been born and raised in a tropical country, who'd never lived in a house with a fireplace, know anything about chimney vents? Or about paraffin logs that smolder with invisible noxious fumes long after the flames go out?

He did take credit for saving us, attributing his ability to stay conscious that night to his lifelong devotion to brandy. "All those years trained me, prepared me," he would say. "If I did not do this"—lifting his snifter to eye level and swirling the Courvoisier—"if I did not know how to be dizzy, I would have fainted like your mother. *Lahat tayo patay.* We'd all be dead."

"Yes, Papa. You being a drunkard saved us."

"Laugh, Mama. It's true."

On this my parents did agree: our family story really could have ended that night. Our American tale could have reached its zenith at the end of six paragraphs on the front page of the local news, with a listing of names and ages and a place to send flowers. My parents woke up to a frightening truth, and even my older brother and I seemed to come into a new awareness of how much *we did not know.* This new land held unseen dangers. You could go to sleep with your Winchester and never wake again. You could pull a lever and lose everything.

My family has turned this story over many times, each of us recalling it a little bit differently. A few of us, notably my mother, saw nothing complicated in the tale. "Your father almost killed us," she'd say, a conclusion that only hardened further after their divorce many years later. He was the one who loved that fireplace, the one who got the logs and lit them up. He was the one who shut the vent.

A couple of us leaned toward a more collective blame. We

thought the incident had more to do with the snow and the celebratory spirit that overtook us all. None of us had ever seen snow up close before. It cast a spell. We all wanted a luminous glow to complete the picture of a white Christmas in America. And besides that, it was bitterly cold — colder than our tropical bodies had ever felt before. We drew close. We all fed the fire.

Over the years I've come to take an even longer view, believing in my most conspiratorial moments that our desire was the culmination of many generations of yearning to be the golden people, Americans, to have their life, to enjoy the scenes that we believed they enjoyed. It might seem a bit of a stretch, but stay with me: the desire traced far into the past. My forebears lived with a terrible love for America, and before that, for Spain. Terrible because it was a love of the conquered for their conquerors.

Lapu Lapu had repelled Magellan, but Spain returned with bigger ships and more powerful armaments, and they took control of the islands just as European colonizers would take control of much of the Asian continent. The British took all of South Asia and eventually subdued the powers of China, the Portuguese and Dutch claimed Indonesia, the French subjugated Indochina, the Russians conquered the vast sprawl of Central Asia and Siberia, and the Americans forced their way into Japan and eventually took the Philippines from Spain. That completed a centuries-long process: Asia as a giant pie sliced up and consumed in any way the European powers saw fit. White masters ruled over yellow and brown multitudes.

To be a Spanish Castilian in Las Islas Filipinas was to be a small god standing over a trembling legion. And when Spain surrendered the Philippines to the United States, the Americans installed themselves as the exemplars, the mini-gods, and the *masa* had no choice but to bow or be silenced. Tens of thousands of Filipinos demanded independence, refused to bow, and were hunted down like rabbits, the survivors left to starve in the hills. No one knows the actual number of Filipino deaths caused by American troops. Some say 200,000. Entire villages and islands were "depopulated." That was the military term: depopulated. The massacre was barely noted in the American history that I learned, and that my forebears learned in their American-run schools. If it was mentioned at all, it was labeled an indigenous uprising. The Philippine Insurrection.

My grandparents bowed to the Americans and sought to learn from them. My parents sought to *be* them. It was part of the grotesque progression. The desire fueled my family's journey across the ocean, leaving everything familiar behind, to plunge into a vast uncertainty with little thought of the perils, the final result of hundreds of years of cumulative reaching for the beloved. The fingers of desire struck the match.

"Your father almost killed us."

Maybe it was that simple. An accident, that's all. Unlikely that a newspaper story would say that a consulate official and his family were almost killed by carbon monoxide fumes caused by centuries of colonial oppression and the subsequent cumulative yearning for an equal share of paradise. But the metaphor has worked for me on another level. My parents risked everything to cross the ocean and live the dreamed-of life, and in their earliest attempt for the perfect moment almost extinguished us all in the process. We have been, unconsciously and not, extinguishing ourselves ever since.

Our early years in America were marked by relentless self-annihilation, though of course we did not see it that way at the time.

Everything was done in the name of love, for the cause of fitting in, making friends, making the grade, landing the job, providing for the future, being good citizens of paradise—all so necessary and proper.

First was the abandonment of our native language and our unquestioned embrace of English, even though for my parents that abandonment meant cutting themselves off from a fluency they would never have again. Possessing a language meant possessing the world expressed in its words. Dispossessing it meant nothing less than the loss of a world and the beginning of bewilderment forever. "Language is the only homeland," said poet Czeslaw Milosz. My parents left the world that created them and now would be beginners for the rest of their lives, mumblers searching for the right word, the proper phrase that approximated what they felt inside. I wonder at the eloquence that must have lived inside them that never found a way out. How much was missed on all sides.

We left behind José Rizal and picked up Mark Twain. We gave up Freddie Aguilar for Frank Sinatra and the Beatles, "Bayan Ko" for "The Star-Spangled Banner" and *She loves you yeah, yeah, yeah.*

My parents' adulation of all things white and Western and their open derision of all things brown or native or Asian was the engine of their self-annihilation. Was it purely coincidence that our first car, first house, first dog in America were white? That our culminating moment in America was a white Christmas? White was the apex of humanity, the farthest point on the evolutionary arc and therefore the closest earthly representation of ultimate truth and beauty.

I grew up hearing my parents' offhanded comments about how strong and capable the Americans were, how worthy of admiration, and conversely how weak and incapable and deserving of mockery their own countrymen were: "They can't do it on their own; they need help." I heard it in their breathless admiration for mestizos—persons of mixed European and Asian blood—how elegant and commanding they were, and the more European the better. To be called mestizo was the ultimate flattery. White spouses were prizes; mestizo babies, blessings; they represented an instant elevation, an infusion of royal blood, the promise of a more gifted life.

One late evening at the White House I was playing on the floor of my parents' bedroom closet, behind a row of shirts, when the door opened. It was my father. Instead of revealing myself, I just sat there watching him in silence, cloaked by a wall of sleeves. He changed into his house clothes and then stood at a small mirror appearing to massage his nose, running an index finger and thumb along the bridge, pinching and pulling it as if to make his nose narrower and longer. He stood there doing that for a short time and then left, shutting the door behind him. I thought it curious but did not think about it again until a few months later, when I saw him do it again as he absently watched television. He didn't know I was in the room.

"What are you doing, Papa?"

It startled him. "Nothing, son. Just massaging."

"Does your nose hurt?"

He looked at me, deciding what to say, and then he seemed to relax. "*Halika dito, anak.* Come here, son. You should do this," he said to me gently. He showed me how to use my fingers to pinch the bridge of my nose and then tug on it in a sustained pull, holding it in place for twenty seconds at a time and then repeating. "You should do this every day. If you do, your nose will become more *tangus*. Sharper. Narrower. You'll look more mestizo. Your nose is so round! And so flat! *Talagang Pilipino!* So Filipino!"

"What's wrong with flat?"

"Nothing is wrong with flat. *Pero* sharper is better. People will treat you better. They'll think you come from a better family. They'll think you're smarter and *mas guapo*, more handsome. *Talaga, anak.* This is true. See my nose? The other day a woman, a *puti*, a white, talked to me in Spanish because she thought I was from Spain. That happens to me. I massage every day. Don't you think I look Castilian?" He turned to show his profile. "*Ay anak.* My son. Believe me."

I did believe him. Just as he had believed his father when the lesson was taught to him decades earlier. These were the givens: Aquiline was better than flat. Long better than wide. Light skin better than dark. Round eyes better than *chinky*. Blue eyes better than brown. Thin lips better than full. Blond better than black. Tall better than short. Big better than small. The formula fated us to lose. We had landed on a continent of Big Everything.

One sunny afternoon, my father and I walked to a hardware store a few blocks from our house. As we were about to go inside, three American men in overalls and T-shirts walked out, filling the doorway and inadvertently blocking our path. They were enormous, all of them well over six feet tall, with beards and beefy arms and legs. My father and I stood looking up at this wall of denim and hair. The Americans appeared ready to scoot over. "Excuse us," my father said, and we moved to the side. One of the men said thanks, another snickered as they passed.

My father leaned down and whispered in my ear, "*Land of the Giants.*" It was the name of a television show my family had

36

started watching, a science-fiction series about a space crew marooned on a planet of gargantuan humans. The crew members were always being picked up by enormous hands and toyed with. The show's tagline: "Mini-people — Playthings in a World of Giant Tormentors." My family was captivated by the show. I think we related to the mini-people who in every episode were confronted by impossibly large humanoids.

Americans did seem to me at times like a different species, one that had evolved over generations into supreme behemoths. Kings in overalls. They were living proof of a basic law of conquest: victors ate better. The first time I sat as a guest at an American dinner table, I could scarcely believe the bounty: a whole huge potato for each of us, a separate plate of vegetables, my own steak. A separate slab of meat just for me! At home, that single slab would have fed my entire family.

The size of American bodies came to represent American capacities in everything we desired: they were smarter, stronger, richer; they lived in comfort and had the surplus to be generous. They knew the way to beauty and bounty because they were already there, filling the entryway with their meaty limbs and boulder heads and big, toothy grins like searchlights, imploring us with their booming voices to come on in. Have a seat at the table! Americans spoke a few decibels louder than we were used to.

We were small in everything. We were poor. I mean pockets-out immigrant poor. We were undernourished and scrawny, our genetics revealing not-so-distant struggles with famine and disease and war. We were inarticulate, our most deeply felt thoughts expressed in halting, heavily accented English, which might have sounded like grunts to Americans, given how frequently we heard "Excuse me?" or "Come again?" or "What?" The quizzical look on their faces as they tried to decipher the alien sounds.

My father, who was a funny, dynamic conversationalist in his own language, a man about Manila, would never be quite so funny or dynamic or quick-witted or agile or confident again. He would always be a small man in America. My mother was small, too, but it was acceptable, even desirable, for women to be small. American men found my mother attractive. She never lacked attention or employment. My father was the one most demoted in the great new land. He was supposed to be the man of the family, and he did not know which levers to pull or push, and he didn't have the luxury of a lifetime, like his children, to learn them.

I'm convinced it was because of a gnawing awareness of his limitations in the land of the giants that he was a dangerous man to belittle. Gentle and gregarious in the company of friends, he was a different person in the larger world of strangers: wary, opaque, tightly coiled. My father stood all of five feet six inches and 150 pounds, every ounce of which could turn maniacal in an instant. He took offense easily and let his fists fly quickly. He was not deterred by mass. He recognized it, but became blind with fury when it trespassed on him or his family. I once watched him scold a man twice his size, an auto mechanic he thought was taking advantage of him, and threaten to leap over the counter to teach him a lesson. "You kick a man in the balls and he's not so big anymore," he once told me. Actually, he told me more than once.

My mother corroborated the stories of my father challenging other men over perceived slights, losing as many fights as he won and getting downright clobbered on a few occasions, once landing in the hospital for a week. My mother was present at some of those fights; she was the cause of at least one, in which an unfortunate young man ogled her and ended up laid out on the sidewalk.

I got another glimpse of his inner maniac once at a park in New Jersey when I was about twelve. A big red-haired kid on a bike spat on me and rode away laughing and making faces. My father followed him all the way back to where his family was picnicking and confronted the three men in the group, all Americans, one of whom was presumably the kid's father. They all appeared startled. I heard only part of the conversation that followed. "We could take care of it right now, right here," my father told the men in a low, threatening voice, his fists clenched into hard knots. He stood leaning forward, unblinking. The men averted their gaze and kept silent. On the walk back to our spot, my father said, "Tell me if that boy comes near you again." I was speechless. His mettle astonished me. But it was something more than bravery on display that day. His fury was outsized, reckless, as if something larger was at stake, and of course now I know there was.

Unlike my father, I worked hard to get along with strangers. We moved so much in those early years that I got used to strangers as companions as we passed from place to place. I learned American English, trained out whatever accent I had inherited, picked up colloquial mannerisms. I kept a confident front, not in a loudmouthed way but in a reserved, alert manner, and I got more surefooted in my interactions as I got better at English. If I had to guess, I'd say my classmates would have described me as a little shy but smart and likable. I brooded in private. How could someone be ashamed and capable at the same time? I was fated to have a secret life.

So I worked on becoming an American, to be in some ways more American than my American friends. But I learned, eventually, that I could never reach the ideal of the beloved. And when the realization came, it seemed to land all at once, blunt force trauma, and I felt embarrassed to have been a believer.

It's one of the beautiful lies of the American Dream: that you can become anything, do anything, accomplish anything, if you want it badly enough and are willing to work for it. Limits are inventions of the timid mind. You've got to believe. All things are possible through properly channeled effort: work, work, work; harder, faster, more! Unleash your potential! Nothing is beyond your reach! Just do it! I believed it all, drank the elixir to the last drop and licked my lips for residue. I put in the time, learned to read and write and speak more capably than my friends and neighbors, followed the rules, did my homework, memorized the tics and slangs and idiosyncrasies of winners and heroes, but I could never be quite as American as they. The lie is a lie only if you fail, and I most certainly did.

When I ask myself now when this shame inside me began, I see that I inherited the beginnings of it from my father, and he from his father, going back in my imagination as far as the arrival of the Spanish ships almost five hundred years ago. An ancient inherited shame. It accompanied us across the ocean. We carried it into a country that told us: not reaching the summit was no one's fault but your own.

The roof of the White House leaked for most of the winter. Water stains on our ceiling appeared like a strange pox. Eventually the cold and wet gave way to a sparkly green spring that transformed our overgrown yard into a lush forest. Our menagerie shrank. My parents gave away the German shepherd when she proved too much dog for my family to tame. Papa released the robins and tried to release the rabbit, but it kept returning to our back porch, still terrified. The parakeets escaped on their own. Our neighbor, the white-haired man who shoveled our sidewalk during the snows, spotted them down the street flitting from tree to tree, chased by crows.

"Should've kept the cage shut," he told us. "Probably dead by now."

My father continued his pugnacious ways into old age. His last fistfight, which occurred when he was sixty-eight, involved a hugely obese teenager who my father believed had disrespected him. The kid, who weighed over two hundred pounds, ended the fight by lying on my father and crushing the air out of him, almost killing him. Papa had had two heart bypass surgeries and a couple of strokes by then and was by all measures frail. He could not concede weakness. In other ways, he did evolve. His gentleness with us, his children, magnified. His remorse for leaving us haunted him and kept him in a perpetual state of apologizing. *I'm sorry, anak. My son. I'm so, so sorry.* He cried easily. And he never again made a fire in the fireplace, in that house or in any other we lived in together.

"Do this every day," he told me.

I followed his advice. As inconspicuously as possible, I hid from the sun. I dangled from tree branches and pull-up bars to stretch my body, praying to gravity, aiming my heels toward the center of the Earth. I tried to eat beyond my appetite, and as a teenager secretly began taking protein supplements to help me grow. I rubbed oils onto my eyelids to keep them supple, to prevent the epicanthic folds from turning my eyes *chinky.* And every night before falling asleep, for at least twenty seconds, I would massage my nose. The shape of your nose determined your fate.

It was the symbol of your lineage, the mark that determined which gate you entered. As I got older, I got more obsessive. I began attaching a clothespin to my nose and leaving it there all night. I was already prone to nosebleeds, and sometimes the clothespin was too tight and I'd wake up with blood on my pillow. To make my lips thinner, more mestizo, I would suck them in and place masking tape over my mouth for hours at a time. Anyone who walked into my bedroom on those nights would have thought I was being held hostage. None of it worked. The mirror mocked me. The clay of my face would never change.

3

Orientals

When you've shot one bird flying, you've shot all birds flying.

—Ernest Hemingway

O ne afternoon in the Bronx, on my way home from school, I crossed paths with a little boy whose face still appears to me forty years later. The first sighting was of his small round head bobbing in the distance. As he got closer I saw he was skipping and spinning wildly, shouting as he moved along. I guessed he was eight or nine. He was wiry with mocha-brown skin and bundled in a junior-sized flight jacket that puffed him up like a pillow. In his hands was a toy tommy gun.

"*Boom-boom-boom! B-b-b-boom!* You're all dead!"

His words pierced the thick afternoon air, turning heads of people near and far, some across the street. The walkers just ahead of him shifted to make room as he approached. "Watch out for this one," a woman just ahead of me muttered to the young girl whose hand she suddenly clasped. The girl drew closer. An older black man in a gray overcoat stopped in his tracks, looking at the boy with great irritation: *Watch where you're going, you little shit. Where's your mother?* The boy shot him. *Boom!* The man shook his head and waved him off. I'd be next. The distance closed between us.

Our neighborhood did not match any picture of America I'd ever seen. My family didn't know before moving here that the South Bronx in the 1970s was in the midst of a long disintegration into a slum. We lived right on the edge of entire blocks already disintegrated. My parents, with four kids, a newborn daughter, and Lola, did not have the luxury to be distracted by such concerns. They awoke each day with the nagging terror that the whole lot of us, with a single missed paycheck, could end up living in a church basement and standing in line at a soup kitchen. Surviving each day was their sole focus. It did not matter to them where we lived because we wouldn't stay long anyway. We U-Hauled it coast to coast twice, seven immigrants, a baby sister, and a dog (my father got another one) folded into a huff-puff car. In twelve grades I attended eight schools. Every move left us feeling more untethered. When people asked where I was from, and there were always askers, I would just mutter the name of the last town we lived in. It was simpler. I was twelve when we moved to the Bronx.

We lived on the Grand Concourse, a block from East Burnside, in a sooty three-story brick building with wrought iron bars over the windows. The back lot was fenced and circled by razor wire. The first two floors housed creaky little offices. A Jewish dentist with terrible teeth, Dr. Hauer, who worked on the second floor just beneath our living room, told us not to bother trying to get rid of the roaches, which had built a thriving civilization in the walls and under the floors. "Look at them like fellow tenants." He smiled. Those teeth. I can still see them.

To get to my school, JHS 79, I walked north on the Grand Concourse, passing grimy storefronts and blocky apartment buildings with darkened windows. I'd cut over to Creston Avenue, a side street with entire blocks that looked like bombed-out ruins: crumbled sidewalks, cars stripped to the axels, the rust bleeding into faded graffiti, abandoned buildings with broken and burned-out windows, dusty lots littered with the detritus of what used to be there, trash everywhere. People lived in buildings that did not seem habitable. During hot months their clothes hung from the skeletons of fire escapes, and kids and dogs cooled off in the brown arc of fire hydrant water spilling forth day and night. In the afternoons, clusters of hard-staring people would sit on concrete steps, smoking weed or cigarettes, some falling-over drunk or strung out — heroin was a big problem — or epically tired and exuding an air of absolute exile that I wouldn't encounter again until many years later in refugee camps. Someone on the steps might offer advice to a new kid on the block: "Yo, how about you give me the money in your pocket so I don't break your arm."

I actually did see a man break a kid's arm for a bus pass. The kid wailed, and no one helped him. Much worse things happened. There were boys at school who just stopped showing up, and I would hear later that they'd been "hurt," the circumstances even more vague. I know for a fact a few had been killed. I remember articles describing the South Bronx as "the worst case of urban blight in the country," "a free-fire zone," "a post-apocalyptic landscape of addicts and thugs." Hollywood made a movie out of it, *Fort Apache, The Bronx.* I remember watching it, seeing the various characters, and thinking, "Hey, I run into people like that all the time."

So encountering sociopaths during my routine comings and goings was not unusual. I learned to avoid eye contact, and that's precisely what I was doing that one afternoon as Gun Boy in the flight jacket whirled toward me. I stared straight ahead, even as his eyes seemed to lock onto me. Suddenly, when he was about to pass, he spun directly into my path, a little tornado, and planted his feet with a thud, the barrel of the gun inches from my face.

"Wanna die, you Chinese fuck?"

I was a foot taller, looking past the gun barrel, down into his face. Round, fleshy cheeks. Smooth brown skin. Lips pressed tight, jaws clenched. Dark, ball-bearing eyes, unblinking. Such concentrated malevolence you would not have thought possible in a little kid.

"Hi there," I said. "What you playing?"

"I'm gonna kill you, you Chinese fuck!" he screamed in my face.

The woman ahead of me stopped to see if a disturbance was unfolding. I waved her away, not taking my eyes off my accoster, this pint-sized thug who, had he been older, bigger, would have been truly frightening. My stomach tightened, and I felt something rising up in me, not fear but the uneasiness that precedes it. I resisted the impulse to punch the kid in the face. (My father had taught me the mechanics of punching at an early age.) My curiosity won out. As I try to remember it now, I see myself cocking my head slightly, in the way of a perplexed dog, and looking to see if there was anything in the boy's eyes I could recognize.

Where did the virulence come from? What tortured circumstances formed him? Was he one of those kids who bathed at fire hydrants, picked through trash for scraps to eat, slept in those burned-out buildings I passed every day on the way to school? I thought of asking if he was okay, but his glare gave no space for that. With a finger I pushed the barrel of the gun to the side, and I met his eyes with my own glare.

"I'm not Chinese," I said. It was all I could muster, a statement of what I was not, which even then I knew was beside the point.

"Chinese fuck!" *Boom. Dead.*

He spun around and continued down the sidewalk, and I walked the rest of the way home as if nothing had happened. And, really, nothing had. New York was full of crazies. Old and young, black, brown, and white, and sometimes tiny.

Later that night the encounter came back to me, swirling in the brew of my thoughts and keeping me awake. What most agitated me was not the belligerence or the gun in my face but that he'd called me Chinese. *I'm not Chinese, you little piece of shit.* He could have called me an ugly fuck or a skinny fuck or any number of other fucks. But to him I was first and foremost Chinese. I wasn't certain that I liked Chinese people, given my dealings with them at school; they seemed clannish and utterly uninterested in me. And I loathed the idea of being indiscriminately lumped in with all the other short, skinny, black-haired people of the world. It was another form of disappearance. I was already working to erase the self I was born with, an effort from the inside. In being lumped in, I felt erased from the outside.

During our first decade in America, we did not experience the violent racism that was my parents' secret fear. Not to say that old-timers didn't think to themselves when they watched our U-Haul pull in, *There goes the goddamn apartment complex.* And who knows what jobs or rentals or loans or cushy assignments we did not get because of some unspoken bias, how often we might have been called simians behind our backs, whether neighbors who smiled in our presence later snickered with their friends. *It's Rosa-velt, dear. Not Roos-belt.* To our faces, anyway, most people were civil if not cordial. Stares we got, occasionally hostile but more often inquisitive. But no rocks through the window in the middle of the night, no crosses burned on the lawn, no death threats left in the mail. No mobs formed to run us out of town. When relatives asked, "How is your life in America?" my father would answer, *Nandito pa rin kami.* We're still here.

We did get called names and the accumulation of epithets flung at us or in some way associated with us over the years — *dogeater, goo-goo, gook, dink, slant, slope, slit-eye, squint-eye, paddy rat, wog, jap, nip, flip, ching-chong, Charlie Chan, chopstick, chop suey, Chinaman, chink* — told me that Americans saw us as part of a larger group with which we did not naturally identify. Americans viewed us through the prism of race, and over time we saw ourselves through the same lens. The group to which we belonged wasn't called "Asian" at the time. That came many years later. The name of our group finally crystallized for me during another encounter on the Grand Concourse.

Many months after Gun Boy, I met Rosemary and Lisa. I'd describe them to my friends later as "two beautiful hippie girls." What made them beautiful was that they paid attention to me. I had just turned thirteen. They were in their early twenties, beads around their necks, free-spirited and talkative in their long loose skirts. Probably college students, now that I think of it (Fordham University was down the road). They both had light skin and white features but looked to be mixed in some way, part black or Hispanic. They had long wavy hair that fell below their shoulders. Leaving their apartment one afternoon, laughing about something, they almost ran me over on the sidewalk, smothering me for an instant in a flowery girl smell. The taller one, Lisa, cupped my face in apology.

"Oh god, Rosemary, look at this adorable Japanese boy!"

"Sweet Jesus, you are so cute. How old are you?" Rosemary said.

I told them my age and grade in school and then stood there like a mute in the midst of two goddesses. They introduced themselves, said they had just moved in and needed to stock up. "We need everything!" They didn't belong in this neighborhood, I thought. They were too open, too indiscriminately affectionate. It occurred to me later, recalling the redness in their eyes, that they could have been stoned. Maybe their free-floating love would have attached to whomever they bumped into; anybody could have been as adorable. Whatever. I was happy to be their favorite human of the moment. Lisa smelled like lilac. So dazed was I that it did not matter that she had called me Japanese.

"Your skin is so beautiful," said Lisa, raising her hand to my face. "May I?"

Okay.

She ran the back of her fingers languorously over my cheekbone. "Feel his skin, Rose. It's unreal."

"Hope you don't mind being molested in public, little boy!" Rosemary said. She took her turn stroking my face. I got a glimpse of a sprout of hair peeking out from her armpit. "Sweet Jesus, can we kidnap you!?"

They laughed.

Lisa tousled my hair. "You are Japanese, aren't you?" she said.

"I bet he's Chinese," Rosemary said.

"I'm Filipino," I said.

"Same thing!" Lisa said. More laughter. I laughed too. It was funny and the lilac was so sweet. "You know what I mean. You're Oriental."

"Yep," I said. "I am."

I knew the word, of course. Years earlier in Seattle, registering me for elementary school, my parents stood at a counter filling out forms. My mother asked my father, "What are we, Papa—Oriental or Pacific Islander?" My father, impatience in his voice, asked the woman behind the counter if it was okay to check both boxes, because, you see, we come from islands in the Pacific Ocean in the part of the world known as the Orient. The woman looked at him blankly. "Go ahead and check Oriental," she said.

I remember the incident more for my father's annoyance. I was seven and thought nothing of the word. It was obviously an acceptable word, official and all that. Proper people used it, people in charge.

Sometime after Lisa and Rosemary, I started a file—one of my first—labeled simply "Orientals." Inside went notes and newspaper or magazine clips, anything that made reference to the word. After a while I had to create sub-files, so large was the universe of things called Oriental: roots, rugs, religions, noodles, hairstyles, hordes, healing arts, herbs and spices, fabrics, medicines, modes of war, types of astronomy, spheres of the globe, schools of philosophical thought, and salads. It applied to men, women, gum, dances, eyes, body types, chicken dishes, societies, civilizations, styles of diplomacy, codes of behavior, fighting arts, sexual proclivities, and a particular kind of mind. Apparently, the Orient produced people with a singular way of thinking. There was no way, wrote Jack London, for a Westerner to plumb the Oriental mind—it was cut from different cloth, functioned in an alien way.

There was a grocery store on East Burnside we called "the Oriental store" where we bought rice. A place near Fordham Road offered "Oriental massage," a few blocks from a shop that sold "Oriental furniture." A travel agency near where my mother worked in Harlem put up posters of the Orient on its walls and windows. My mother and I wandered in there a few times. The posters and pamphlets showed images of geishas and monks and fog-shrouded temples. Of elephants in gold jewelry. Of strange-looking boats in dark waters, open markets teeming with ant-like hordes in strange dress, women with baskets on their heads, children in paddy hats sitting on the backs of hulking black water buffaloes. Of farmers in pointy hats turned down toward the earth like rows of bent nails, and rice terraces wrapped around the foothills of green volcanoes still steaming from their last eruption.

What a mysterious place, the Orient. Menacing but also immensely alluring. So otherworldly. So alien, as London said. No wonder it spawned people with such unfathomable minds. I had read *White Fang* and *Call of the Wild;* Jack London was one of my favorite writers.

To know why the word "Oriental" chafes so many of us today, it helps to know its history. The word came from the Latin word *oriens,* meaning east or "the direction of the rising sun." The Romans named the eastern part of their empire Praefectura Praetorio Orientis, which included the eastern Balkans and what is now Syria. The Western understanding of the Orient expanded eastward as Western explorers went deeper into Asia until such time as Europeans used the word to describe the vast stretch of the planet east of themselves all the way to the Pacific Ocean.

The Orient came to encompass a quarter of the globe, including Egypt, Nepal, and Korea; Turkey, Mongolia, and Indonesia; Lebanon, India, and Japan.

Europeans popularized the concept of the Orient at a time when they were usurping much of it. Colonizers used scholarly studies on the Oriental mind, Oriental character, and Oriental society as guides to subduing and managing their subjects. Concept and conquest went hand in hand.

The underlying assumption of Orientalism was that the Orient represented the inferior opposite of Europe: the East was feminine and passive, the West masculine and dominating. The East was spiritual and inward-looking, the West rational and outward-seeing. The East was bound in tradition, the West impelled by progress. The East was primitive, vulgar, and defenseless; the West was the beacon of civilization, the standard of refinement, and the wielder of unstoppable military power. The Orient needed to be civilized for its own good.

Aside from being backward, we Orientals were also cravenly submissive, incurably exotic (from the Greek *exotikos*, meaning "from the outside"), inscrutable, cunning, silently treacherous, and highly penetrable. In fact, begging to be penetrated. We Orientals lived to be acted upon by virile, dynamic, rational Westerners.

Eventually, the Orient came to refer most commonly to what's now East and Southeast Asia—China, Japan, Korea, Mongolia, Taiwan, Vietnam, Thailand, Singapore, the Philippines, Malaysia, Laos, Indonesia, Cambodia, Myanmar, and Brunei. The people here—in contrast to those from West, Central, or South Asia—were the most easterly of Easterners, the yellowest of the yellow race.

• • •

Yellow was the perfect color for Orientals. It was only superficially descriptive of skin tone. The cultural associations with the color resonated with the Western view of the Orient. *Caucasoids,* or Europeans, were white, the color of purity and power. *Negroids,* or Africans, were black, for their dark and animalistic character. *Mongoloids,* or Orientals, were yellow, the color of infirmity and cowardice.

Orientals, without knowing the Western associations, did not object to their assigned color. Yellow had an altogether different history in the East, a regal history in fact. The ancient Chinese divided the spectrum into five "pure" colors, with yellow representing soil or "of the Earth." This interpretation may have been born in northern China, where sediment deposits from the Gobi Desert every year turned the rolling plains a deep golden color. Depending on the context, yellow covered a spectrum of hues from light beige to gold to orange to reddish-brown. The 3,400-mile waterway said to be the cradle of Chinese civilization is called the Yellow River. Yellow became the color of royalty. And its favored rank spread to other countries. The color came to symbolize bravery in Japan after the 1357 War of Dynasties, during which warriors wore yellow chrysanthemums as pledges of courage. Hindus in India wore yellow to celebrate the Festival of Spring. The primary symbol of the Philippine flag is a golden yellow sun, signifying a new beginning. Today, yellow is the color of unity among Filipinos.

In the West, yellow's unsavory status went back centuries. Judas Iscariot, the betrayer of Jesus, came to be associated with yellow (although there's no link in the Bible), and the color grew to represent envy, jealousy, and duplicity. In France, the homes of traitors were painted with yellow shellac. The color also came to be associated with illness. In the Middle Ages, the body was thought to contain four distinct fluids—blood, phlegm, black bile, and yellow bile. Yellow bile supposedly made one "peevish, choleric and irascible." Yellow fever caused jaundice, a liver condition characterized by a yellowish discoloration of the skin and eyes. Illness brought frailty in body and spirit. In nineteenth-century America, to be called yellow meant you were cowardly.

Yellow was a contemptible color in the Western imagination, and it fit snugly with the Western idea of the Oriental human being, especially as he became a threat first to the American workingman, then to the chaste American woman, and finally to Western civilization itself. Alarmists called for the white race to brace for the onslaught: all of Christendom would be overrun by the Orient's yellow swarms! They would conquer not by might but by sheer number, a human tidal wave. Pulp-fiction authors fed the hysteria with descriptions of slant-eyed immigrants practicing heathen religions, raping white women, and dancing on the ruins of white civilization.

The late nineteenth and early twentieth centuries saw a determined effort by whites to prevent yellows already here from multiplying—blocking entry of yellow women and banning marriage with whites. In the western territories came a furious drive by whites to get rid of yellows altogether. An uncounted number of Chinese were lynched, villages were purged, entire settlements wiped out. The hatred fueled an anti-Oriental fever that swept up yellows of all nationalities. The vision of "the menace from the East was always more racial than national," writes historian John Dower in *War Without Mercy*. "It derived not from concern with any one country or people in particular, but from a vague and ominous sense of the vast, faceless, nameless yellow horde."

The "Yellow Peril" became a persistent theme in American

politics and culture through World War II, when the term was applied to Japanese, those treacherous simians who snuck up on Pearl Harbor, and who made it necessary to round up and lock up the Japanese in America. Something like 110,000 American citizens of Japanese ancestry were forced from their peaceful lives and herded into internment camps. By this time, white Americans had passed a series of laws preventing more yellows from coming over, and those restrictions stayed in place until the civil rights movement forced lawmakers to rethink the country's immigration policies. Orientals in large numbers (equivalent to the numbers coming from Europe) could not legally immigrate to the United States until the 1960s, the decade when my family arrived.

I was ten when the My Lai Massacre in Vietnam became news. I remember coming upon the *Life* magazine pictorial of the killings, so many golden-brown women and children and old men, as many as five hundred, shot up and contorted in bloody piles along dirt paths. Bullet holes like giant sores on legs and arms and necks. Bodies that had come apart. And the faces: open-mouthed, sometimes with eyes still frozen in terror, brains spilled into black hair. The faces looked like those of my family. My aunts and uncles, my sisters. Oriental faces. I looked at the photographs a long time, could not stop thinking about them.

The only American soldier convicted in the killings, Lieutenant William Calley, served three months under house arrest. What the massacre drove home to me was that Oriental life was not terribly valuable. You could extinguish hundreds of Orientals—unarmed villagers, farmers, women, toddlers, infants—and the penalty would be napping and watching television in your apartment for twelve weeks. I still have that *Life* pictorial in my files and run across it once in a while. The same emotions well up each time.

Having met other immigrants like myself in America, I can say that a great number of us came to our same "Oriental" identity in a similar fashion. We arrived in the United States as Japanese or Korean or Filipino, but over time we became Orientals. It wouldn't be until the 1970s, after Edward Said's book *Orientalism* shook up the academy and garnered an influential following, that "Oriental" began its descent into scholarly opprobrium, along the same path as "Negro" and "Indian."

Nevertheless, many older Americans still use the word, often innocently. In the Midwest and South, I'm frequently referred to as Oriental by kind and well-meaning people. I can walk a few blocks from my house in the Pacific Northwest and order an Oriental salad and an Oriental chicken sandwich. (I've been tempted to ask for an Occidental beverage to go with them.) The word is still used by some who know its associations and accept them. A Filipino colleague of my father's, in a mock-Japanese accent, told me, "If you eyes rike dis," using his fingers to narrow his eyes, "you Oriental. If you eyes rike dis," pulling them wide open, "you da boss."

But in academic and government circles, "Asian" became the correct designation, and I became one in college. All of us former Orientals were now officially and properly Asians.

I took on the new designation at a time when immigrants from Asia were entering the United States in unprecedented numbers. When my family arrived, fewer than a million Asians lived in America. In the 1960s, the U.S. government finally acknowledged its racist policies and opened the gates. Over the next couple of decades, 3.5 million Asians moved to the United

States, constituting the second great wave of immigration from the Big Continent. The second-wavers were more diverse. Among them were Indians, Koreans, Vietnamese, Cambodians, Laotians, Hmong, and Mien. Many came from war-ravaged countries. Very many came with nothing.

As a journalist in my twenties and thirties, I wrote extensively about these communities. No surprise, I found each group exuberantly complex and distinct, and perceiving themselves as separate from—and often antipathetic to—other Asian ethnicities. The parents and grandparents clove to their countrymen, the Vietnamese with other Vietnamese, Koreans with Koreans, Cambodians with Cambodians.

It was the children and grandchildren, the ones growing up in America, who would find—or be coerced onto—common ground. Years of checking "Asian" on countless forms, of being subjected to the same epithets and compliments, of living in the same neighborhoods and housing projects, and sharing similar challenges and aspirations—the most important to become Americanized—all of these would compel young Vietnamese, Cambodians, and Filipinos to accept their belonging to the category known as Asians.

Perhaps the most unifying force was the perception that everyday Americans saw them as the same, and what made them the same was their "racial uniform," to use a term coined by sociologist Robert Park. The uniform was thought to consist of a certain eye and nose shape, hair and skin color, and body type, usually shorter and skinnier—identifiers of the Yellow or Mongoloid or Oriental and finally now the Asian race.

Young people who would have had no natural ties in Asia found themselves bound together in America, and more so with succeeding generations. The farther out in time from the point of arrival, the more Asian they became. It mirrored what happened to Africans brought to America as slaves. "We may have all come on different ships," Martin Luther King Jr. said, "but we're in the same boat now." We Asians were now in the same boat. Our uniform did not lie. Like Lisa said on the Grand Concourse: Japanese, Chinese, Filipino — same thing!

A news item out of Detroit in the summer of 1982 — the murder of Vincent Chin — made an impact on many of us. Chin was the adopted son of a Chinese laundry owner. He was twenty-seven, a Motor City native, and about to get married. He and some friends celebrated his bachelor's party at the Fancy Pants strip club in Highland Park, not far from where he grew up. As they cavorted with the dancers, two white men at a nearby table — Ronald Ebens and stepson Michael Nitz — began making racial remarks. Among other things, the two repeatedly called Chin a "nip" (a derogatory name for Japanese), and one declared loudly, "It's because of little motherfuckers like you that we're out of work."

Ebens was a plant superintendent for Chrysler, Nitz an unemployed auto worker. Massive layoffs in the auto industry were being blamed on the phenomenal success of Japanese imports. Ebens's boss, Chrysler chairman Lee Iacocca, quipped that the solution might be to nuke Japan again. Michigan congressman John Dingell decried "those little yellow men" taking jobs from hardworking Americans. The Yellow Peril resurrected. Writer Helen Zia was a laid-off Chrysler worker living in Detroit at the time. She recalled the climate in the city:

Local unions sponsored sledgehammer events giving frustrated workers a chance to smash Japanese cars for a dollar a swing. Japanese cars were vandalized and their owners were shot at on the freeways. On TV, radio and the local street corner, anti-Japanese slurs were commonplace. Asian American employees of auto companies were warned not to go onto the factory floor because angry workers might hurt them if they were thought to be Japanese.

At the Fancy Pants, Ebens and Nitz took their frustration out on Chin. They provoked a fight, and the club threw out both groups. The father and stepson retrieved a Louisville Slugger baseball bat from their car and tracked down Chin outside a nearby McDonald's. They knocked him to the ground and Nitz pinned his arms while Ebens swung the bat again and again, caving in Chin's skull. The groom-to-be died four days later. A local boy who held two jobs, Chin was as hardworking as any American. But to his attackers, he fit the bill. Anger at the Japanese spilled over onto anyone who *looked* Japanese.

The story became a thread that joined Asian American groups with little previous contact. Help was offered, partnerships were formed, and compacts made among Chinese, Japanese, Taiwanese, Koreans, and Filipinos. These formal ties mirrored the informal ties coalescing in streets and schoolyards. A new pan-Asian awareness seemed to come into being. Some scholars consider Chin's murder a watershed moment, the event that gave rise to Asian America as a social entity. Behind it all was a keen awareness among Asians that it could have been any one of us at the McDonald's that night.

Chin's murder signaled something else, a shift in the perception of Asians and Asian Americans as a threat from above rather than below, and this shift in status seemed to correlate with the adoption of "Asian" to replace "Oriental." The colonial use of "Oriental" assumed an inherent inferiority in the people of the East. If Orientals posed a threat to Westerners, it was because they could take away low-skill, low-level jobs from white laborers, or they could contaminate white racial purity by intermarrying. The threat came from below because Orientals were seen as lower in the hierarchy.

But in the 1980s a new fear arose: perhaps these Asians were actually *superior*. Not in their biology (although some posited that), but in their culture. Japan had risen in a few short decades from cataclysmic defeat to become the second-largest national economy in the world. Its cars were taking over America's roads. China had begun its spectacular economic ascent, and Taiwan and South Korea were on the move. Pundits observed that Asians might in fact be smarter, or in any case they appeared to be more industrious, more disciplined, more driven, and more willing to sacrifice for long-term gain. Gore Vidal, in *The Nation*, predicted the emergence of a new global order with "yellows" at the top. If whites failed to meet the challenge, Vidal warned, "we are going to end up as farmers — or worse, mere entertainment — for more than one billion grimly efficient Asiatics."

In the United States, Asians were considered a "model minority," a notion that has persisted into the twenty-first century. Asian immigrants were seen to be more studious, more entrepreneurial, more civic-minded, and more apt to create cohesive communities. The success fomented a new kind of resentment. Like Chin, Asians were sometimes viewed as still somehow connected to a vast foreign empire that increasingly stole jobs from "real Americans," and not just manual jobs anymore but positions in engineering, health care, skilled manufacturing, and communication. Or they were seen as taking over enrollment at elite universities or dominating entire job sectors such as computer programming and technology—to the point where many observers suspected the existence of invisible "caps" to limit the number of Asians. The "bamboo ceiling."

"Asian" came with a new set of stereotypes attached, many of them reflecting Vidal's fear of a cold, machine-like efficiency. Asians today are often seen as "tech geeks," "math wizards," "book nerds"—small, shy, studious moles who could gradually and ever so inconspicuously come to dominate the world.

We lived in the Bronx for three and a half years, and I never saw Gun Boy again. He still occasionally skips and twirls through my mind. I still see his eyes, black ball bearings encased in brown flesh. An encounter that lasted thirty seconds has stayed alive in me for four decades. I believe it's because he stepped on a newly exposed nerve in me. The question big Joe Webb asked me in that clangy stairwell at JHS 79 became a repeating echo.

"What you supposed to be, motherfukka?"

What, in fact, *was* I supposed to be? I tried on various uniforms of manhood in my teens and twenties. I was a white suburban frat boy for a while. I was a cool ghetto nigger for a very short while, complete with a salon Afro and pick. I was a swivel-hipped Latin lover, a Puerto Rican hipster, a Mexican *cholo,* a Native American soothsayer. I looked to Crazy Horse and Chief Joseph for guidance and for a while sought visions like they did in their last days before annihilation. I slipped in and out of these uniforms like someone in a fitting room at Macy's trying on suits, except I'd walk around in them for a few months or years. The only suits I avoided were Asian ones. I had already concluded that being Asian did not play in one's favor as a man, not in this land of giants.

I ran into Lisa a few more times before we moved. She was as gregarious as ever, laughing over every little thing. It was a fake laugh, but I didn't care. Once at a pizza stand on Jerome Avenue, Lisa bought me a Sicilian and we talked like friends—"So how's school?" she asked—but I noticed she only half listened to my answers. Her eyes scanned the sidewalk for people she might know. I wished I were more interesting, older, taller. I inhaled her lilac fragrance and would try to keep the memory of it until I lay alone in my bed at night. I imagined being with her. My brain had been increasingly overtaken by sex and romance, usually in that order. More than anything, I looked forward to being in union with a girl.

Once I saw Lisa walking with a small group, holding hands with a young scraggly-haired man with a sparse beard. He wore a poncho. I craned my neck to get a better look, to see what kind of man Lisa had chosen, but I lost them in the moving flock of bodies. I never saw her again after that. For the next year, until we moved out of the Bronx, I would pass her apartment building on the way to and from school, slowing my pace, lingering as I approached the entrance, and after passing it, looking back to see if the doors opened behind me.

4

Seeking Hot Asian Babes

Black hair
Tangled in a thousand strands.

— *Yosano Akiko*

can still picture her, those glue-on lashes fluttering over a baby face. We did not exchange a word, never acknowledged each other except for a moment when our eyes met as she walked past me. Barely a fleeting glance. Yet this girl, who was trying so hard to look like a woman, comes to mind whenever I recall that first trip to Cebu. I had been shaken by what I saw during my various excursions around the island, but one recurrent sight stood out.

More than the Magellan monument, the reason for my trip. More than the armies of child beggars that came from nowhere to tap on the windows at every major intersection, their fingers like hard rain, their faces desperate and cunning at the same time, filthy, gap-toothed. More than the shantytowns that clung to riverbanks sliding into oblivion. More than the barefoot crowds gazing up at the giant billboard faces of Richard Gere and his Pretty Woman. More than my driver Bobby asking me, "Want girl tonight? I can get. One hour, no problem. Two hours, whole night, easy. You like virgin? I can get."

More than any of these things, I was struck by the girls and young women who walked in a sweethearts' clasp with white men old enough to be their fathers or grandfathers, or even, in some cases, great-grandfathers. I had read about this, was aware I might encounter it, but I was not prepared to see these couples so close, in the flesh, and so many of them. They were everywhere, flitting in and out of sight, stepping from taxis, disappearing into crowds, into clouds of exhaust, into the lobbies of hotels and their unlit corridors. My eyes followed them.

"You like?" Bobby would ask. "I can get."

No. Just drive.

"Yes, sir. Maybe later. No problem."

One night at a dinner buffet on Mango Avenue, I was seated next to a cluster of potted shrubs, and just on the other side sat one of these couples, no more than a jeepney-length away. I made it a point to sit facing in their direction. She looked to be fifteen or sixteen, short and the slightest bit pudgy in her tight dress and spike heels. The heavy makeup seemed only to accentuate the baby fat of her face. The blood-red polish on her nails looked to have been applied in haste. He had silver hair and a pale dewlapped face that reminded me of an old shar-pei dog. His light-colored eyes shifted under prickly white eyebrows. I guessed he was in his late fifties or early sixties. He was the only white person in the room. The couple sat mostly in silence, each with a hand on the table, hers caressing his.

"Want more noodles, darling?" she cooed.

"I'm fine, I'm fine," he said, glancing at me. I looked away and then picked up the newspaper I had brought with me.

"Are you sure? I can get," she insisted, her eyes shining with devotion.

"Okay, okay," he said. "Maybe a little more. Not the skinny kind. The fat ones," he said with an edge of impatience. His accent sounded American, upper midwestern. Minnesota? Wisconsin? He worked a kink out of his neck. "I don't like the skinny ones, okay."

"Not the *bihon*, the *Canton*, right, darling?"

"Yes, get me the fat ones," he said. "Not too much. You always get too much. Half portion only, you hear me?"

She caressed his hand in a slow circle and then gave it an affectionate pat before rising. "Of course, darling, the *Canton*. You love the *Canton*."

He grabbed an empty glass and thrust it at her. "You can get me some more of this while you're at it, okay."

She took the glass and kissed him on the cheek before turning toward the buffet, which sat in my direction on the other side of the room. I watched the glow from her eyes quickly fade and her lips straighten into a grim line, as if the strings holding up her face had gone slack. In the moment she passed me, her expression looked more like those of the children on the street tapping on car windows.

I set the newspaper down and glared at Mr. Shar-pei, my head cocked accusingly. I *wanted* him to see me this time, and our eyes met for an instant. He quickly turned away, clenching and unclenching his jaw.

I looked to the other dinner guests, to the hostess behind her podium, to the slick-haired manager in the suit whispering instructions to a busboy. The screaming thought in my head:

Is *anybody* going to do anything about this? I imagined the reverse scenario—a sixty-year-old Asian man walking into a crowded restaurant in Boise and smooching with a fifteen-year-old blonde in a tight dress and heels. Wouldn't someone act? Maybe I was ascribing to Americans a capacity for indignation that did not exist, that I imagined for this occasion so I could feel self-righteously disgusted at the locals who let Mr. Shar-peis do anything they want. My indignation must have seemed naïve to the old desk clerk at my hotel, to whom I reported the incident. He told me that's what happens on these islands: foreigners do whatever they want, and locals smile and call them "sir."

"You'll get used to it," he said.

Whenever I am tempted to think that Asian women fared better than Asian men in the colonial enterprise, I try to remember Mr. Shar-pei.

I was new to Asia at the time. The buffet scene was the first of many similar sights and scenarios that I would see play out all over the continent: foreign men with money and power flying in and out on jumbo jets, local women trapped and destitute in a way most Westerners could not grasp.

Mud-hut destitute. Dollar-a-day, fried-rice-morning-and-night, shit-in-a-bucket destitute. Poverty so dismal that peasant women find it acceptable to copulate with a dog on a small stage six nights a week as drunken men holler for more. To let multiple fat foreigners penetrate all their orifices at once with Trojan-sleeved cocks and any phallic object in reach, for the price of a small thin-crust pizza. To lick the balls of foreigners with their butter stink and doughy porcine thighs, and to pamper them like sultans—*More Canton, darling?*—so that maybe, just maybe, if the slots line up and love develops, these men might consider bringing them home to the US of A. Jackpot!

It's not just American men. They come from all over western Europe, Australia, and New Zealand. From the oil-rich nations of the Middle East. From the privileged mestizo classes of Latin America. They drive the sex tourism industry in cities like Angeles and Olongapo in the Philippines; Svay Pak village in Phnom Penh, Cambodia; the Patpong district of Bangkok, Thailand; and other, smaller and lesser-known places like Mango Avenue in Cebu. In Thailand's Isaan region, an estimated 15 percent of marriages involve young Thai women with Western men in their sixties.

American writer Boye De Mente, in his dated but still revealing book *Bachelor's Japan*, offers his thoughts on why Western men are attracted to the Asian woman: "She has an innocent-appearing, baby-faced cuteness that is particularly appealing to Western men (because the average Western woman is far from cute in this way and it suggests youth, innocence, etc.). She is also small-bodied and delicate-appearing, like a young Caucasian girl of fourteen or fifteen years of age, and therefore has the appearance of forbidden fruit that is, however, accessible." Plus, De Mente says, Asian women offer an added benefit for men insecure about their sexuality. These men "lose this sense of fear because *they feel inherently superior to them* and therefore need feel no shame."

I've seen Boye De Mentes and Mr. Shar-peis and their Asian girls strolling beneath the walls of light on Nanjiing Dong Lu in Shanghai, a city turned by Western colonials into "the whore of Asia." I've walked a dusty stretch of Jalan Pasar Kembang Street in Yogyakarta, Indonesia, and watched laughing foreigners window-shop for girls as if shopping for shoes or baseball gloves. I've stood outside the Hotel Kunlun in Beijing, ten minutes from Tiananmen Square, and seen the line of taxis delivering powder-coated escorts with their stilettos and fake Guccis to prim businessmen checking their watches in the lobby. I sipped espresso with a call girl in Singapore who said to me in a refined British accent, "What's the matter, dear boy, you're acting jealous. You want us to scurry for you, too?"

Of course, Asian men have always partaken in the for-sale pleasures available in their countries. As Asian economies rise, so do the numbers of men — from India, China, Japan, Taiwan, South Korea, and other ascendant nations — who travel to neighboring countries in search of sex for sale. And it's almost always men — local, Asian — who pimp out the women and profit most from the trade. Nevertheless, the sex-tourism industry, particularly in Southeast Asia, grew into its modern incarnation largely to fulfill the desires of Western men.

In Thailand, for example, there were ten thousand to twenty thousand prostitutes in 1957. Then came U.S. involvement in the Vietnam War, during which American soldiers used Thailand as a staging area for military operations and, more important, as a getaway place for rest and recreation. At the height of the war, the number of Thai prostitutes soared to several hundred thousand. Now the number on any given day, according to various estimates, ranges from 300,000 to as many as 2 million. Of the 20 million tourists who travel to Thailand each year, 60 percent are male; of those males, as many as 70 percent come specifically for sex.

The West won, after all. And Europe's drive into Asia had always been fueled in part by sex. Along with the quest for fortune and adventure, the promise of erotic encounter, fed by whispers that fanned into legends, kept explorers and adventurers and eventually soldiers and armies moving toward the eastern horizon.

It may have begun with the ancient Greeks. Their goddess of love, beauty, and sexuality, Aphrodite, was an amalgam of goddesses whose origins traced to "the Orient," now the Middle East. Aphrodite was an early embodiment of the desirable Eastern woman. Her cult of worshippers planted the seedling idea of the Orient as female, decadent, and seductive.

In Homer's *Iliad*, two Greek alpha males, Achilles and Agamemnon, vie for a young female slave, Briseis, a captive princess from Asia Minor, introducing the idea of the Eastern woman as a spoil of Western victory. After the Roman conquest of Jerusalem, the emperor Vespasian had special coins made. One side showed a triumphant Roman soldier, "muscular and manly"; the other depicted a woman—representing Judaea—with her head bowed in submission, suggestive of "impending violation," writes Richard Bernstein in his book *The East, the West, and Sex.*

Perhaps more than anyone else, the Venetian explorer Marco Polo inflated the rumors to mythic proportions. Today, many scholars question his veracity. But starting in the thirteenth century, he worked up legions of Western men with his narratives of vast harems and free-flowing sex, which he claimed to have encountered during his travels to the East. The heads of households along the Silk Road, Polo reported, "give positive orders to their wives, daughters, sisters, and other female relations, to indulge their guests' every wish." He described the great Kublai Khan gathering hundreds of beautiful maidens and doing "with them as he likes." Of Chinese courtesans, Polo wrote that strangers who sample their charms "remain in a state of fascination and become so enchanted by their wanton arts that they can never forget the impression." Other explorers and narratives would follow. Tall tales of silk-haired maidens with horizontal vaginas and oversized clitorises inspired legions of European men to see for themselves. The Oriental woman became the object of mystical fascination.

Kings and princes across Europe read Polo's narratives. Christopher Columbus was greatly inspired, so much so that Polo's writings accompanied him across the Atlantic as he searched for an oceanic route to the East. They influenced Chaucer and Dante. And other Western writers over time — Flaubert, Melville, Gide, Conrad, Maugham — expanded upon the idea of the Orient as a vast sexual playground.

There was a measure of truth in their narratives. "Harem culture," as Bernstein calls it, was created by Eastern men for their own benefit and existed long before Europeans came along. This culture ensured that there were always groups of women reserved exclusively for the sexual gratification of privileged men. These women included the khan's and the emperor's concubines, the sultan's cloistered virgins, the *datu*'s mistresses, the general's field wives, the shogun's geisha, the chieftain's maidens of the night — women who were at best subordinate companions, at worst chattel.

Concubines could be sold, traded, given away, abandoned, or killed at the pleasure of their masters. The ancient Chinese general Sun Tzu once famously ordered the public beheading of two of the emperor's favorite concubines just to make a point (the women had giggled at an inopportune moment). Concubines were buried alive with their newly dead master to keep him company in the afterlife. Women of the East lived and died in a man's world long before Westerners came and conquered.

The West's piecemeal conquest of the East spanned four centuries, a male enterprise to its molecular core: conceived and carried out by men, for men, in the ravenous spirit of men. Colonization, said British scholar Ronald Hyam, turned "the whole world into the White Man's brothel." In Asia, the templates were already there, and in many cases it was just a matter of stepping in and assuming the master's role.

This happened as recently as World War II, when American occupying forces took over the use of Japan's sex slaves. Japan had forced eighty thousand women from the conquered nations of China, Korea, and the Philippines to provide sex to Japanese soldiers. These "comfort women" endured the unimaginable. At war's end, the United States had the run of Japan, and according to testimonies of surviving women, Americans made full use of the "comfort stations." Control over the women was simply transferred from one master to another.

Over time, harem culture would transfigure into a capitalistic enterprise in which Westerners' buying power gave them untrammeled access to whatever carnal pleasure they desired. Anything could be purchased, including sex with children. Today, humanitarian groups estimate that as many as a third of all prostitutes in Southeast Asia are under eighteen. If the East made women and children slaves, the West turned them into penny commodities.

The same class of women who were concubines in earlier eras now perform sex work. The sex worker in Asia goes by numerous names and occupies many shadow vocations: hostess, companion, escort, personal attendant, entertainer, girlfriend, masseuse, bar girl, bath girl, dancer, mistress, maid. I've been to beauty salons in Manila, in which the female workers cut hair, do manicures, and provide sex on demand. In parts of China, they're referred to as *er nai,* meaning "second breast." In Japan, sex workers engage in *baishun,* which roughly translates to "selling youth"; or *enjo kosai,* "compensated dating"; or *oderibarii herusu,* "health delivery." The Korean term for sex work is *yun-lak,* or "ruining the ethics."

In the Philippines, Vietnam, Cambodia, and Thailand, many sex workers think of their chosen path as their last best hope for the dreamed-of life. Through a human rights group based in the Philippines called GABRIELA, I met a number of women who worked as short-term "companions" for visiting foreigners. I spent time with these women and their families. The hope of nearly every last one was that love would strike and produce a ticket to the West.

I guessed the young woman with Mr. Shar-pei was such a companion. "Of course, darling, the *Canton.* You love the *Canton,*" she said, playing him for the ultimate prize. Mr. Shar-pei had the wallet. He held the key and the ticket to paradise. He was her knight in shining armor. The young woman must have known her chances were slim, that it was far more likely that Mr. Shar-pei, at the end of his vacation, would board the jumbo jet home and never be heard from again.

Sometimes, though, the men do bring the women home.

In the winter of 1994, a burly forty-seven-year-old American named Timothy Blackwell brought a young Filipina, Susana Remerata, home to Seattle. He was a handyman, overweight, balding, and by most accounts awkward with women. She was twenty-one when they first made contact. She was pretty and petite, barely one hundred pounds, and preoccupied by dreams of going to America.

The two met through a matchmaking agency called Asian

Encounters, which had provided Blackwell with a catalog of Filipino women. The cover promised "Gorgeous Pacific Women," "Pearls of the Orient: Ladies Known for Their Beauty, Charm, Grace and Hospitality." Blackwell selected two dozen women and purchased their names and addresses from the agency. He contacted them all. Susana's enthusiasm put her at the top of the list.

Blackwell traveled to the Philippines to meet her and was treated like a VIP in her village, Cataingan, a community of farmers and fishermen on the island of Masbate. He strutted his beefy frame through town, bigger than anyone else, and bathed in the aura of glamour associated with white Westerners, particularly Americans. Blackwell played the part. He lavished gifts on Susana and her family, his dollars representing wealth beyond anything she had known. Susana had grown up in a house with no electricity or indoor plumbing.

"It's him. He's the one!" she told friends.

Blackwell would eventually pay for a grand wedding. The whole village attended, or it seemed that way to him. He tracked his expenses meticulously. He claimed to have spent $10,000, including the cost of Susana's airfare to Seattle. Susana got her wish. She made it to the land of dreams. But Blackwell was not what he pretended to be in Susana's village. She was shocked to learn that he was of limited means, a fact that became clear the moment she walked into his tiny apartment. The American segment of their union lasted thirteen days.

In court she claimed he was abusive, manhandling and bruising her, scolding her like a child. He described her as cold and calculating, using him to gain U.S. citizenship. He tried to have her deported. She voiced her suspicion that he was a homosexual. They were in the final stage of an annulment trial when, in a court hallway, Blackwell calmly pulled a nine-millimeter pistol from his briefcase and shot Susana three times, so close the gun barrel almost touched her body as it slumped on a wooden bench. He was convicted of murder and sentenced to life in prison. Susana's body was shipped back to Cataingan and buried in a graveyard just a bicycle ride away from where she was born.

I covered the story for the *Seattle Times* and later worked on a segment for *60 Minutes*. The reporting revealed a world largely unknown to me: the business of matching First World men with Third World women. The power imbalance that I had glimpsed on Mango Avenue with Mr. Shar-pei and his teenage companion was, in this business, formalized, officially built in to the transaction. Many of the men count on the inequality. The "Pearls of the Orient" were expected to submit to the men, to serve them with no demands of their own. It's what Oriental girls are supposed to do, what the catalogs implicitly promise. If the women did not meet expectations, the men could send them back, like Blackwell tried to do with Susana.

"You go through a couple of divorces in this country, and you get a bellyful of American women and their liberated ways," a middle-aged man from Everett, Washington, told me. He was a long-haul truck driver who had also found a wife through Asian Encounters. His bride met expectations. "Now, you go to the Philippines, to Asia, and those girls have a serving attitude. Everyone does. It's part of the culture. They believe the man is the head of the home."

Starting in the mid-1990s, online services began replacing printed catalogs. Through these services, thousands of Asian women each year leave their home countries—four thousand from Cambodia, seven thousand from the Philippines, ten thousand from Vietnam — as fiancées or wives of foreigners. In some cases, it's more convenient for foreign men to move to their bride's home country, where foreign currency goes a lot farther. For instance, Thailand's northeastern region is now the permanent residence of an estimated 100,000 foreign husbands. In some villages, 80 to 90 percent of husbands are German, Swiss, or American.

The websites often advertise the women just as the old print catalogs did, banking on the Oriental girl mystique. A Google search using the words "seeking Asian women" pulls up 24 million results, and lists an astounding number of websites offering ways for Western men to hook up with Asian babes — abroad or in town, for love and marriage, for a summer fling or a sixty-minute Oriental fantasy at the local Motel 6.

Many of the sites link to pornography. A common theme is the sexual domination by "enormous" white or black men of "tiny" Asian women. Asymmetry is a significant part of the appeal. "Asian" is a separate genre in Western porn, and it refers almost exclusively to Asian women. Asian-ness in porn often means an inordinate amount of humiliation for women.

In 2001 I covered a story in Spokane about the abduction of two Japanese girls by a trio of locals who, it came to light, had an "Asian fetish." The term is a colloquialism referring to a strong-to-obsessive sexual interest in Asian women by non-Asian men, usually white. Roger Bragdon, Spokane's police chief at the time, called it "one of the most despicable crimes" he'd ever dealt with. Bragdon's wife was Japanese American, and he seemed to take the crime personally, assigning an unprecedented twelve detectives to the case.

The perpetrators were members of a local sadomasochist sex club who had cased a Japanese school on the edge of town, the

American branch of Mukogawa Women's University. It was on seventy-two wooded acres above the Spokane River. On a cold November morning, the trio kidnapped the two teenagers from a bus stop and took them to a house in the Spokane Valley where they tortured and raped them for seven hours before releasing them. They videotaped the assaults. The ringleader was forty-year-old Edmund "Eddie" Ball, described by a friend as "a huggable, muscular, tattooed, evil-looking teddy bear who carries a large Bowie knife and a bullwhip and whose hobby happens to be torturing people."

One of the girls, so traumatized she could hardly speak, became suicidal. The other girl wanted to return to Japan immediately, but stayed in town long enough to help police track down her abductors. In their confessions, the assailants admitted two previous and unsuccessful attempts to abduct Asian-looking girls in the Spokane area. They were planning another abduction at the time of their arrest.

Ball was found to have a large collection of Japanese bondage videos, some obtained during a trip he'd taken to Japan four years earlier. I could easily imagine him prowling Tokyo's red-light district, Kabukicho, and cultivating his fetish. I could see him peeping into those darkened rooms where fake samurai force geishas to submit, walking into those shops that rent life-sized Japanese sex dolls with functioning orifices, slipping into those reconstructed subway cars designed for paying customers to grope women dressed as schoolgirls. I could see Ball spending lots of time in those subways.

He walked the Tokyo streets just as Timothy Blackwell had done in Cataingan and Mr. Shar-pei in Cebu: surveying the possibilities and seeing the local women through the lens of their fantasies. Just as tens of thousands of Western men, and an increasing number of newly rich Easterners, do every year. One way or another, they bring home images and impressions, tall tales and short ones, rumors that grow into legends; they write travelogues and blog posts; they whisper to friends as Ball did to his accomplices, and the stories feed the mythology. *Little brown fucking machines!*

In America, the mythology had its own beginnings. Immigration polices in the nineteenth century allowed Asian men to come as laborers, without their women. Americans feared that letting both sexes in would spawn yellow hordes. The limited number of Asian women permitted entry were often picture brides, war brides, mail-order brides, or prostitutes. An element of sex was a prerequisite for admission. Chinese women in certain western towns were immediately presumed to be whores. These women reinforced the European notions of the Orient, notions that accompanied European immigrants across the Atlantic and white pioneers across the American West — right into the modern day. Oriental girls existed to be consumed by men.

I've seen up close the effects of this mythology on Asian women in America. Life for them is trickier to navigate, more fraught with unseen pitfalls. Jackals and opportunists lie in wait. Every accomplishment comes with doubt and self-questioning and no small amount of apprehension.

"I'm always nervous about the moment when the other person realizes I'm not what he thinks," a Chinese American woman in New York told me. "They like you when you act the way they think you should act. When you don't, it's a big disappointment. Or they turn on you. The whole stereotype can kind of backfire, you know? It's like, 'Oh, I'm supposed to be sweet and submissive? Sorry!' Then all of a sudden they act like you're a dragon bitch."

Which brings up the other, opposite archetype of the Asian female in the Western imagination: the Dragon Lady, a domineering, imperious, coldhearted woman willing to do whatever it takes to get her way. So she is either a docile lotus flower or a flame-breathing dominatrix.

I knew a woman at work who went from flower to dragon. She was Chinese American, slim and bright. As a new hire, she was received by senior editors in the newsroom as a darling, a sweetheart, an ingenue who needed tutelage. But as time passed, she revealed herself to be adept and, in some instances, formidable. As she rose in rank, the attitude among many of my colleagues, both men and women, moved to the other extreme. She became relentlessly gossiped about as ruthless and cunning. A devious megalomaniac. "A snake," one colleague said, and I can still hear the low-throated rancor in his voice. The shift was startling.

The same dynamic applies to Asian women whose face or figure does not conform to stereotypes, and as the woman from New York told me, "That's most of us." An Asian woman who looks nothing like a China doll or geisha not only gets ignored but is made to feel hideous. New York described a friend from Taiwan who was bigger boned and chubby in the face, and who never seemed to have any luck on the dating scene. In high school, she was called Miss Piggy. New York grimaced. "Basically, if you don't look like the fantasy, you're uglier than ugly."

And this from a young Vietnamese American woman in Los Angeles: "I always have to wonder if the guy who's staring at me, or the guy chatting me up, or the man interviewing me for a job, if they're really seeing *me*. Or are they playing some porno in their head and I'm the Oriental slut?"

"I had a guy come up to me at a club," she continued. "I was with some girlfriends, and this white guy comes up all smooth-talking and the love in his eyes, and he says, 'I hear Asian girls are wet all the time.' I mean, come on. It wasn't his opening line, but it came up fast. I'm thinking 'So suave. Go away.' I told my girlfriends. We laughed, but it creeped us out. They had their own stories. It made us all want to go home."

5

Babes, Continued

Baby I can't please you, I can't please you, oh baby.

— *Sam Phillips*

Women everywhere in all times have been more vulnerable than men. They've been held down and kept down, and have suffered immeasurably more at the hands of their opposites. This is true of women of the East and Eastern women in the West. But I also believe that, in certain spheres of life in twenty-first-century America, Asian women have it better than Asian men.

The women from New York and Los Angeles, whom I mentioned earlier, admitted to me that no matter how nettlesome and reductive the Western stereotype of Asian women, it's still preferable to the Western stereotype of Asian men. "Oh, no question," said New York, who immediately apologized. Why preferable? Because given the choice between being sought after or ignored, most people would choose the first, even if the attraction is based on superficiality or falsehood. Even if it brings the occasional creep or the occasional awkward moment. It's better than being not regarded at all.

The mythology born from Orientalism says that Asians make up a feminine race, with both genders a little farther along the scale toward the soft extreme. The same racial uniform that emasculates the men feminizes and eroticizes the women. If the men are the least manly of men, the women are the most womanly of women — more pliant, more sensual, more sensitive and attentive to the needs of the stronger sex. This perception often works in favor of Asian women in the areas of love and courtship in modern America, in part because Asian women seem to have retained their allure without the colonial inferiority that once came with it.

"Hot damn, I love Asian women!" a college friend used to tell me. He ogled them shamelessly, and courted a few doggedly. A senior news editor, a middle-aged white man who was otherwise one of the most politically correct people I knew, once told me the same thing in almost the same words, minus the exclamation point. I have heard the words many other times in various permutations, and they are usually uttered without hesitation, because they reflect the culturally sanctioned notion that Asian women are *desirable*.

Jerry Seinfeld in a 1994 episode of his show tells his friend Elaine that he should have pursued the woman lawyer Donna Chang. Why? Because, Jerry says, he *loves* Chinese women. When Elaine asks whether that isn't racist, Jerry says if he likes their race, how can it be racist? To make the same point, the show could have used a woman of any Asian nationality. I hear the echoes of my Grand Concourse friends Lisa and Rosemary:

Chinese, Japanese, Korean — same thing! Vietnamese, Thai, Filipino — vixens all!

On college campuses, Asian women have been publicly declared hot commodities by fellow students. At New York University, a 2011 story on the school's news blog — titled "'Yellow Fever' Is Causing a Stir at NYU" — received 708 Facebook "likes," twenty-one Twitter tweets, and seventy-three comments, many quite lengthy. The "fever" refers to the phenomenon of white males seeking Asian females — sometimes obsessively — for love and sex. The story's author, May Wang of NYUlocal.com, reports that the fever "is taking this country by storm." The two Asian females interviewed say they don't mind the attention from white men; they prefer white men. "Bring on the hot Asian fetish, boys!" one says. "There are enough of us to go around!" The one white male in the story admits to liking the dominant role in relationships and says he "absolutely" has "yellow fever." He says it's "because of the kind of personality Asian girls have" and "their petite builds." The lone Asian male laments that it doesn't work the other way around: "Unfortunately, Asian men can't get white girls!"

The Harvard Business School's student newspaper, the *Harbus*, came out with a similar story a few years earlier titled "Sex and the Campus: Attack of Yellow Fever." A story published on the West Coast, which opens by quoting a student at the University of California at Irvine, was headlined "Yellow Fever: They Got It Bad, and That Ain't Good." These stories — interestingly, all written by Asian women — report on a phenomenon that residents of large metro areas have observed for many years.

Dating studies find that most whites in the United States prefer the romantic company of other whites. But of the white men open to dating outside their group, a significant number find Asian women attractive. And it's mutual: many Asian women openly seek romantic unions with white men. Women of all races consistently rate Asian men least attractive. A two-year study by Columbia University found that "even Asian women find white, black, and Hispanic men to be more attractive than Asian men." Along with black women, Asian men occupy the bottom rung on the desirability ladder, and there is considerable lamenting from both groups that their opposites have run off en masse with white partners.

The dating patterns spill over into marriage. Nuptials between white men and Asian women occur more often than any other interracial combination. Up to half of Asian-origin women in the United States marry white men. The out-marriage rate among Japanese American women is as high as 80 percent. White male–Asian female couples outnumber Asian male–white female couples by at least three to one by conservative estimates and as much as twenty to one in some regions.

But I did not need these statistics. All I had to do was turn on the television or open a pop-culture magazine. They remind me that high-profile Asian American women in the United States tend to partner with white men. Among them: authors Amy Tan and Maxine Hong Kingston; broadcasters Connie Chung, Ann Curry, and Emerald Yeh; athletes Michelle Kwan and Kristi Yamaguchi; actresses Lucy Liu, Sandra Oh, and Kelly Hu; comedian Margaret Cho, fashion designer Vera Wang, artist Maya Lin, commentator Michelle Malkin, former U.S. labor secretary Elaine Chao. There have also been a slew of marriages between high-profile white men and much younger Asian women, starting with Woody Allen and Soon-Yi Previn, the adopted daughter of his former partner Mia Farrow. Media mogul Rupert Murdoch coupled with Wendi Deng, network chief Leslie Moonves with news anchor Julie Chen, actor Nicolas Cage with twenty-something Alice Kim, billionaire George Soros with violinist Jennifer Chun, and producer Brian Grazer with pianist Chau-Giang Thi Nguyen.

Actually, I did not even have to turn on the TV. I could just wander around the households of my own family members. I ended up with six sisters after my father remarried and started a new family. Of my five married sisters, four exchanged vows with white men. My only single sister unabashedly admits her desire for a white partner. I certainly could not blame my brothers-in-law—all fine men—for choosing my sisters, who are, to the very last, smart, beautiful women of substance.

No doubt most of these women, my sisters included, were active participants in the selection process; they probably chose as much as they were chosen. But when Asian women with white partners, including my sisters, tell me race had nothing to do with their mutual attraction, that their love was blind to race and color, I can only fall silent. Whatever I said could only be a form of sour grapes, right? I believe in their sincerity, but it's difficult from my standpoint to accept that race played absolutely no role.

Asian women and Asian men in America, in some ways, live on such different planes that discussions on the topic of cross-race attraction often degenerate into name-calling. What may feel like an innocent "preference" to Asian women sometimes feels like utter betrayal to Asian men. Asian-male lamenting can sound like character assassination to Asian women. We're all influenced by forces larger than any one of us. The women I've mentioned, besides heeding their preferences, were following their instincts for survival and improving their lives, and instincts have a power difficult to resist. Instincts are shaped to an immense degree by the culture we live in, and it's often futile to blame people for absorbing the values of their culture. The most I could do is suggest that other values may exist just outside the boundaries of our awareness.

Being desirable brings advantages well beyond dating and marriage. Numerous studies in the past three decades have found—and, really, poets have always known—that attractive people, at least on the earthly plane (as opposed to the spiritual one), live more blessed lives: they get more attention and encouragement, mate earlier and more frequently, earn more, climb faster, receive more opportunities to advance, and much more often get the benefit of the doubt when it counts.

My mother and father landed in America together but lived out opposite realities in the land of dreams. From their first days here, my mother was granted more choices than my father. She was embraced by white America; my father not as much. In her journals, my mother routinely remarked on how "welcoming" and "generous" her American supervisors were. She'd note that so-and-so offered her a job, asked her to lunch, found a special chair for her aching back, gave her a sought-after shift and, behold, even a promotion. Mr. Fill-in-the-Blank kept her informed of upcoming opportunities, recommended her to superiors, brought her to gatherings of movers and shakers, and so on, month after month, year after year.

Her benefactors were almost always white men. I knew some of them well. They became friends of the family. They adored my mother. Her career trajectory was a slow but steady climb. By early middle age, she was a high-ranking physician at a hospital where she became a beloved figure. My mother was easy to love. She was caring and conscientious, and willing to put in the necessary hours to get done what needed to be done. She probably would have found success anywhere she went, in any vocation she chose.

My father's trajectory went in the other direction. He left Manila with a law degree and great ambition, but after years of going nowhere became a subsistence real estate broker in Portland, Oregon. He had a few good years — with mostly Asian clients — but was never able to sustain anything financially. He was always broke. He grumbled about white Americans looking down on him. My mother's journals chronicled his long list of disappointments: jobs he almost got, promotions he was denied, projects that fell through, ideas that never found traction. My father struggled to earn a livelihood to the end. He never found a groove. It was in large part because of his inability to cope with my mother's success in the face of his perceived failure that their marriage crumbled. In his mind, the word "failure" was written across his forehead.

I'm using a narrow lens for the purpose of this discussion. A plethora of factors contributed to my parents' disparate fates, and their dashed marriage. But it was evident to me, even at an early age, that my mother received favored status in the New World, passed through open doors where my father found them locked. She had crossover appeal in white America. Appeal led to access led to opportunity. My father did not get all those green lights to cross over.

The pattern played out in the most visible branch of my chosen profession. Early in my journalism career I was struck by the number of female Asian news anchors on television, and by the corresponding scarcity of male Asian anchors. These were among the most coveted and highest-paid jobs in the news business. The women were always—*always*, no exaggeration here—paired with white male anchors. The national prototype was Connie Chung with Dan Rather on the *CBS Evening News* in the mid-1990s. Today, you can hardly travel to any metropolitan area in America and not find an Asian American woman anchoring some segment of the local news. In Seattle, there are Asian women anchors on all three network-affiliated stations.

Television places a premium on physical attractiveness—telegenic appeal—and Asian women and men, as we know, are perceived to occupy opposite ends of the spectrum. But you could never get TV news executives to say this publicly. The explanation from the suits would run along the lines of what Stephen Tschida of WJLA-TV, the ABC affiliate that serves Washington, D.C., was told early in his career—that he simply did not have "the proper look" for a television anchor. It remains to be seen whether the widespread acceptance of Asian women in anchoring roles will lead to executive posts with real decision-making power in news organizations.

As a reporter covering new immigrants from Southeast Asia in the 1980s and 1990s, I observed the pattern again among young people: the girls generally did better in school, were more sought after and active in community events, were more likely to prosper; the boys were more likely to lag behind, drop out, and get into trouble. There were successes and failures among both genders, naturally, but the picture repeated itself often enough to be discernible and disturbing.

I recall sitting in a living room with members of a Cambodian family I had been tracking for a story. The parents, new to English, let their two teenaged children do the talking. The sister and brother sat at opposite ends of a peeling vinyl couch. The sister was seventeen, gregarious and smartly dressed. She exuded an air of someone accustomed to being liked, and in fact she was popular at school and proud of her boyfriend, who was white and also popular. The brother was almost sixteen, dressed in jeans and tank top, with a shaved head, a blank expression, and a mouth that barely moved when he spoke. His words came out like grunts. He hadn't attended school in over a year, and he had just been released from juvenile detention after serving time for assault. He was cautious with me, and there was a hardness in his eyes that you would associate with someone much older.

The contrast was startling. They shared the same parents, same upbringing, same housing project, and for years even the same bunk bed — sister on top, brother on the bottom. How did they end up so different? I may have found some clues. One day when the subject turned to why he'd dropped out of school, the brother said one reason was that he didn't like the students, especially the girls.

"Stuck-up bitches," he said.

"Because they don't like you," the sister said. "You and your friends, puny Asian boys trying to act all street. Why would they even look at you?"

"You're just like them," he said.

"Jealous little Asian boy trying to act all street."

Maybe this was only sibling sniping. Maybe the sister-brother contrast was based in inherited traits or cultural influences; maybe one was a family favorite; maybe one and not the other had experienced trauma. None or all of the above? I could not know. Privately I was tempted to overlay on their situation the same theory of unseen social forces that shaped the fates of my mother and father. I could imagine the bud of the same dynamic beginning to flower in that stuffy little living room.

Puny Asian boys, the sister said. She was otherwise a likable young woman, but when she said that, I could feel the old congealed wounds crack open. I was one of those boys. Very few things hurt my young ego more than an Asian female openly shaming me for *my* Asian-ness. If she could not accept me, who could? If even Asian women saw the men of their own blood as less than other men, what was the use in arguing otherwise?

It's not an uncommon sentiment among Westernized Asian women. According to the Columbia University dating study, "even Asian women find white, black and Hispanic men to be more attractive than Asian men." Though I already knew this, I reread the sentence a few times, still wishing it wasn't true. Internet videos and discussions tend to use plainer language. Threads on interracial dating often include Asian women expressing outright disdain for the men who share their ancestry.

"Last week an Asian guy asked me out," says comedian Esther Ku, a Korean American, in a much-viewed YouTube video. "I thought, like, 'When are they gonna realize Asian girls are just way out of their league?'"

In online forums, posts like this one from a Chinese woman are not uncommon: "All women like strong men. White men are stronger than Asian men. When an Asian man stands next to a white man, he looks like a chicken — small, short, less muscular, unconfident. White men are freer. They have better lives. Asian men work too much and show too much respect for authority." Or like this one, from a writer named Jenny An, in the online publication *XOJane:* "I'm an Asian girl. I don't date Asian guys.

Yep, I'm one of those that date lots and lots of (mostly, but not always) white guys. Why? It's simple: I'm a racist." She goes on to write: "I still see myself as a minority. And with that, pretty soon comes connotations of 'outsider.' And I don't like that." Dating Asian men, she writes, would ostracize her into "an Asian ghetto." Dating white men, on the other hand, "means acceptance into American culture. White culture." An's August 2012 essay elicited nearly 1,900 comments, and was widely discussed in Asian American forums.

A YouTube video titled "Why Asian Girls Go for White Guys," produced by a University of California, Berkeley, student, has been viewed over 6 million times. It features a cast of young Asian American women offering descriptions of white men as tall, beautiful, masculine, and commanding to corresponding descriptions of Asian men as short, ugly, effeminate, and insecure. One woman refers to an Asian male friend as someone who could easily be mistaken for a girl. "I can see why white girls wouldn't go for Asian guys," says another with just a hint of a smirk. By the same token, one interviewee isn't surprised that white men pursue *her*. "Who doesn't like Asian girls?" she says, beaming a winsome smile. "Everyone likes Asian girls!"

I had a crush on a Chinese American girl in college, Leny. She was a year older, from a small town in California. She had an open, angular face, a nice figure that she knew how to accentuate, and a surfer-girl sunniness that put everybody around her at ease. I liked her physical dimensions, how they complemented mine. I thought we looked good together. We both came from big Asian families, grew up around whites, and ended up at a state university in a very white state. She was also trying to figure out how her Chinese roots fit into the who-am-I puzzle, a subject we rarely talked about directly because it seemed to make her uncomfortable. We skirted the edges of it.

She may have been aware of my crush, but it did not matter because she dated only white guys. She made that clear early on. The sole reason she allowed me in her orbit was that I reminded her of one of her brothers, and she confided things a sister might tell only a sibling. Our talks would often turn to who she was dating, who she was breaking up with, who was trying to ask her out. Her love life was marked by high turnover. For a while, a couple of Asian guys from Carson Hall were circling her airspace, testing for possible approaches.

Leny held them at bay. "I mean, would *you* date one of those guys?" she said to me. It was her answer to my suggestion that she give them a chance. "No thanks." She liked big guys, she said. Physically big. Socially big. "You know, outgoing." She had no interest in timid men. I'd find myself standing straighter in her company, speaking with a little more volume. None of it mattered. Years later, long after we lost touch, I had a dream about her.

It's a windy day and we're walking across a footbridge over a whitewater river—the bridge to Autzen Stadium. Leny is confiding something. Our hands keep brushing. Suddenly there's a gust, and she grabs my arm, her face open, her eyes searching. An intense longing to kiss her comes over me. Then I see something in her eyes. What is it? A small point, like the tip of a pencil. A dot of hardness in the center of the circle of her pupils. I turn away. The sound of the river disappears. Leny disappears. I walk off, fighting the urge to turn back, to look for her. I keep walking. Because even in a dream, I know nothing hurts more than being rejected by someone who should love you.

6

Asian Boy

Pretty girls make graves.

— *Jack Kerouac*

I came to live in Eugene, Oregon, by way of the slightest of reasons. My family left the Bronx in the summer of 1974 as my parents' marriage was coming apart, and we U-Hauled it once more across the continent, a tension-filled journey that we all hoped would end in peace and our finally planting roots. We stayed two years in the pit stop town of Umatilla, Oregon, whose entrance sign read POPULATION 750. PLEASE DRIVE CAREFULLY. My parents soon ended their twenty-five-year union, leaving my siblings and me devastated and lost in this strange world of sagebrush and hedgehog cactus.

We were more fractured than we could understand. Our known universe had ceased to exist, and we floated through the months like debris in space, utterly abandoned. My father, who had brought us to America, wrote a long letter to his children, apologizing for his failure and pledging his undying love, and then left us to start another life.

My mother moved us farther west, to the capital city of Salem, where I graduated from high school with a class of strangers. I overheard a couple of those strangers talking about the University of Oregon, a leafy green campus wedged in a valley between the Pacific Ocean and the Cascade Mountains. A smart, progressive place, they said. A school where young people studied to be journalists and activists, and virginal coeds abandoned their inhibitions and ran wild through the forests. That sounded good to me. My college search began and ended in twenty minutes. Eugene was an hour south of our house. Financial aid and a $100 monthly allowance from my long-suffering mother set me free, and south I went on Interstate 5, without the slightest clue about what to study or where it would lead.

I was seventeen. My body elongated, grew new hairs, became possessed by new urges and new anxieties. I could no longer pretend to be a boy. But manhood proved elusive—the idea of it, how it was supposed to manifest in a body and face like mine. What did manhood look like, sound like? How was a man supposed to carry himself? I found no footing, partly because I had no male figure to show the way. My older brother was caught up in his own quest for belonging. My father in his last years with the family had become lost to himself, receding into a private life that none of us knew about at the time. He had gone emotionally AWOL long before he physically left. My twelve grades of education, instilling the idea that men like me served mainly as props in the human drama, only added to my angst.

In college, I got used to being the only Asian in class. There were students from Taiwan and Japan and the Philippines, and several dozen from Hawaii, but on a campus of twenty thousand, and in a city of a hundred thousand, we were dots in a white sea. And I was dully aware that most everyone around me knew the order of things, knew who mattered and who didn't. Knowing they knew made me feel less *seen*, even to myself. It was a sensation of dissolving into a barely discernible human form, a shadow of a shadow.

My way of coping was to act the opposite of how I felt. To speak louder than the timid voice I was born with. To walk and talk and act bigger than my five feet seven inches and 155 pounds. To be more colorful than the gray personality I had come to know as myself. It was all contrivance. I took a weight-lifting class, found that I liked it, and continued for several years, packing a layer of muscle over my slender frame. I practiced with collegiate wrestlers and learned takedowns and holds that could choke a man to unconsciousness. I joined the university karate club and refined my striking skills, learning how to hit *through*, rather than *at*, a target. Once, a friend at the dorm dared me to hit a concrete wall, and I did it without hesitation, using all my training to hit a spot just beyond the wall. *Crack.*

"Jesus! You okay!?" he said.

"Of course," I said, as if I did it every day. I smiled like an idiot and went to the hospital the next day. The X-rays showed two fractured knuckles and tissue damage in my wrist.

The masquerade worked to an extent, and some days I forgot it *was* a masquerade. For a while, during most of my twenties and early thirties, I imagined myself as tough. Wolverine tough; ferocity in a small package. Some days I felt my father's fire and itched for someone to push me so I could take him down. It happened a few times, and I did indeed take down a couple of bigger men. In retrospect, I was lucky not to have encountered anyone truly dangerous. My bravado could have gotten me killed. I had skills but not the bone density or mass, and my knees were as brittle as teacakes. Anyone could have downed me with a tap to the side of the knee. But the main reason I *couldn't* be genuinely tough was that I really didn't want to hurt anyone in a serious way. I could sweep a man's legs from under him, but I did not have the killer instinct, like my father, to gouge an eye or cave in a trachea.

Still, the guys in my dorm respected me, and a few girls showed interest but mostly wanted to be "just friends."

I made a good friend to women because I listened as if I really cared, and it wasn't an act. I liked everything about women, and I liked them in every shape and color. I loved their silhouettes and smells and ways of moving. Their general softness, their skin and hair, fingers and toes and everything between. Their necks and lips and eyelashes. Their voices, the sounds they made. Nothing enlivened me more than the company of women. Nothing made me want more to seize the day. To be strong for the moment. To be a soldier on guard, a protector. To be an entertainer. A bard. If I were a poet, I would write only about women. Only women could make me want to *be* a poet. They were better humans, in my opinion. The affection wasn't mutual. My sense was that female eyes did not see me. Even when looking right at me, they seemed to see through me.

I came to learn that in places where everybody seemed on the hunt — bars, nightclubs, bookstores, supermarkets, parks and beaches, football games, social functions of all kinds — selections were made quickly, instinctively, narrowed to a few candidates within moments. It never felt as though I made the cut. Not once in my years at the university did that magical meeting of eyes across a room happen to me, as it did in the movies, as it did to some of my white and black friends. It wasn't for lack of trying. I was Mr. Laser Eyes. But I was not exempt from the curse. Conversations with friends — male and female, Asian, white, and black — fed the dread, confirmed the suspicion that I belonged to an undesirable category.

On the one hand, it offered a clear explanation: the reason women did not desire me was that I was undesirable. On the other hand, it killed me. I was *undesirable*. Between puberty and marriage, there probably isn't a characterization that cuts deeper. It goes to the marrow, finalizes the transaction. It does not have the venom of insult; I could have taken that, could have responded in kind. It is more a declaration, a dispassionate statement of fact: a bank teller saying you have insufficient funds. There's no good retort to that.

"There's something about them," was my Chinese American friend Leny's explanation for why she did not date Asian men. "I mean, would you date one? Honestly, would you?"

"They're kind of repulsive," my white friend Christopher told me more than once. "No offense."

"They need to drink more milk," my black friend Jenny said.

"It's hard to know what's going on behind those little eyes," said Jeff, a white guy who claimed to be part Navajo and who prided himself on being progressive and inclusive.

I had become so Americanized — whitewashed — that my college friends would claim to forget I was Asian. It was the reason they felt free to say these things to me. They saw me as one of them. "You're not Asian. You're Alex" was how Leny explained it. "Shit, man, I don't think of you as a my-nority," Christopher liked to say. I was so lonesome in those days that I was grateful to be part of one club, at least. Belonging somewhere felt nice, and it allowed me to entertain the illusion that I was different from the other Asian guys on campus with their books and lonely stares. But as soon as I stepped away from my circle, I quickly turned into just another Asian cipher.

I realize now how profoundly absorbed I was in my self-loathing, so much so that I failed to see the other exiles around me who must have felt equally unseen. The overweight, the shy, the awkward. The non–English speakers, the weak, the slow-witted, the too-smart-for-their-own-good, the poor. The handicapped, the traumatized, the alienated, the self-exiled; philosophy majors, almost everyone in creative writing, the entire math department. We could have commiserated, formed a communion of the wretched and brought out the diamond in one another. But I did not make room for them in my psyche. They put too inconvenient a wrinkle in that smooth soft blanket of self-pity in which I had wrapped myself.

My Korean friend Kim was another wrinkle. He did not conform so well to my narrative of inferiority. I met him in the weight room at McArthur Court my freshman year; he was a senior. Born in Korea but raised in the States, he spoke perfect English. I learned from classmates that he'd been a star high school athlete somewhere in the Midwest. He was all muscle and jawbone, tall with heavy shoulders and the rounded glutes of a football player. He may have been the first Asian man ever to model Oregon Ducks running wear in the campus newspaper, the *Daily Emerald*. Whenever I went to the gym, Kim would come up to me with his meaty hand outstretched for a firm shake, a broad smile across his face.

"How are you doing, my friend?" he'd say.

He made a point to touch base with every Asian, male and female, who entered the gym. As if to welcome us, to reassure us that we had a protector in that dungeon of clanging metal and grunting white and black beasts.

The only way to break free of the stigma of our category was to be exceptional in some way, and Kim was an exceptional physical specimen, not to mention well spoken and congenial. There were a few like him around campus, Asian standouts who exuded a quiet confidence. The ordinary majority of us could only marvel and shuffle back to our dorm rooms. No amount of grunting in the weight room could turn me into Kim.

I knew more Asian guys like Cho. He introduced himself as Joe around campus, but I learned his real first name was Cho, and it was the name I called him just to be aggravating. He was a Chinese American from Honolulu. I met him one night standing solo by the bar at a basement frat party. He seemed so separate from the scene. Music blared and bodies bounced and whooped it up, the cement floor sticky with beer, a mirror ball casting flecks of light on transported faces. Cho's was the single glum countenance in the room. He looked the way I felt but didn't dare show. He was the only other Asian guy there. He had long silky black hair parted in the middle, round fleshy cheeks, and eyes that seemed coated in armor. He absently held a cup in his hand.

"You tending the bar?" I said.

"They only have beer," he said tersely. His hair fluttered when he turned his head. We chitchatted for just a minute before he turned and said, "This is not against you, but I gotta get out." And he did, making a beeline for the stairs, hair swishing behind him.

Later that night I ran into him coming out of a 7-Eleven with a couple of friends. "You again!" I said. He was a different person out there, loose, rambunctious. He said there were "too many *haole* snobs" at the party. The four of us ended up at a bar guzzling beers, and four glasses turned to eight and then twelve, then a pitcher and fries, and after his friends left, it was just Cho and me and a whole lot of built-up resentment spilling forth. He had the same demons as I; the difference was Cho let them speak out loud.

"So there's this drop-dead chick in my buddy's dorm room," he was saying, holding a single fry like a pointer. He swayed slightly in his chair. "Brunette, big eyes, fine ass, and I'm thinking, 'Yeah, I'm going for it.' I say to my buddy, 'Brah, why not introduce me to your friend.' He says, 'Joe, my man, you know I would but I can't.' I say, 'Why not?' He pulls me into the hall, says, 'She goes out with black guys.' I say, 'Yeah, so?' He says, 'I mean she *only* goes out with black guys. Sorry, dude.' And I'm thinking, 'What the fuck! You got girls who only do white guys. You got girls who only do black guys. Where are the chicks who do Asian guys? Where are they?'" Cho shook the fry before popping it in his mouth. "Where the fuck are *they!?*"

On the street, two young men on scooters buzzed past, narrowly missing a bicyclist with no lights going in the opposite direction. They each turned back to glare at the other. On the opposite corner, a lone woman in coveralls emptied a trash bin into a large rolling cart, her face hidden in the dark. She moved methodically, slowly, in the shadows. I wondered if anyone had spoken to her that evening, if anyone ever spoke to her, or if all her nights were spent in a silent shadow world.

"Asia," I said.

"What?"

"There're probably lots of them in Asia."

Cho cocked his head to the side. "How does that help me, you asshole?"

• • •

I did have a few brief romances in college, two of them involving young white women who — like Lisa and Rosemary — were initially drawn to me because I was different from anybody they had known. I was exotic. There were worse things to be. I played up my difference when it suited the occasion — for laughs, for sympathy, for whatever I could get. I was a racial opportunist. In classes, I would introduce myself the way alcoholics introduce themselves at AA meetings: "My name is Alex Tizon, and I'm an Asian male." I'd refer to myself as the token Asian boyfriend and pull out my handy-dandy calculator as a prop.

I was acutely self-conscious, and it was my way of breaking the tension I imagined around me. I wanted to get the race thing out of the way, to preempt whispers. And perhaps, now that I think of it, to communicate to others my awareness that I was different from them and would never be like them. Somehow I thought if I made that clear, everybody would relax just a little bit. I wanted them to feel comfortable even if I did not, another mode of self-annihilation. I don't know if it had the desired effect on anybody. I was a fool on many fronts.

During those stretches when I forgot about my Asian-ness, those days and weeks when I would slip into a blithe unknowing, somebody would say something to snap me out of it. A beautiful girl named Rebecca, about whom I conjured many a fantasy, telling me in the most sincere way, "I think you're nice-looking, you know, for your race." Her quizzical look at my expression, the corners of my mouth not knowing which way to turn.

Being exotic never lasted. My white girlfriends came to realize, after the novelty wore off, that I wasn't that different after all. It worked both ways. Like men of color around the world, I was hypnotized by the image of white women as the ideal of beauty.

Once I peeked under the skirt of that mythology, the spell began to dissolve. Race played a role in our coming together but was irrelevant in our coming apart. We broke up for the most mundane reasons.

For a couple of months I dated a young black woman, Charmaine, who attended a community college in Salem. She played basketball and was a couple of inches taller than me. She had a lovely face and the most quintessentially feminine figure of anyone I had ever put my hands on. I could almost wrap my fingers around her waist. Our main connection was a shared sense of feeling out of place. Salem was even whiter than Eugene, and we both wondered about the cosmic reasons for our winding up where we did. Our common status made talk easier. It was a relief not to have to explain everything. She understood why I hated Hop Sing and Kwai Chang Caine. She knew the paradox of being stared at and not seen. She knew what it felt like to walk out of a movie theater feeling ashamed or erased.

Alas, like exoticism, exile was a quick-bonding glue that did not last. After intoxication came the letdown, and our first letdown began when she took me to church. Within a half hour, maybe fifteen minutes, it dawned on me that we were foreign to each other in that way. She felt it too. For one thing, Charmaine spoke in tongues and believed I should as well. Long story short, she met someone who had the right tongue, and I went my way.

There were others during my early twenties, including a young woman from the Midwest whom I met in a tiny fishing village in Alaska. She stands out in my memory for a singular reason, and it was that she loved fellatio and liked to provide commentary afterward. So my young adulthood was not devoid of romantic contact, but these encounters were specks of dust in space: random, fleeting, disconnected. In between them were vast expanses of nothing, which, when I recall them even now, so many decades later, bring an ache of loneliness in the middle of my body.

I also see that some of my isolation was self-imposed. That the story I came to tell about myself wasn't as neatly wrapped in rejection as I often liked to believe. When I force myself to dig deep into memory, into cracks and crannies so easily over-looked, I can find instances when women did extend themselves to me, a few in very direct ways. I was the one who said "no thank you." I was the one who turned away because I felt no attraction, or convinced myself I didn't. I was afraid. Perhaps I would disappoint. Perhaps I would find a genuine affection, and then what would I do with my defining narrative of being less than? Who would I be then? I sensed that I did not have enough substance to absorb another's full attention, to actually receive what I wanted most.

There's probably a name for all this, a syndrome of some kind. I've heard the term "self-selection" applied to these symp-toms. Selection in the Darwinian sense, as in "A lot of Asian guys select themselves out of the game." As in "You Asian dudes are selecting yourselves to extinction!" The idea is that in this survival-of-the-fittest society, some Eastern men in the West, often without knowing it, choose to reject themselves before anyone else can. We preempt defeat by taking ourselves out of the game. Whatever the name, I have detected it in other Asian American men throughout my adult life. It reassured me on one level that my observations were not all imagined. That unseen forces created a common experience for men like me.

I sat on the front porch of a house in Seattle one night with two other Asian guys as a raucous party took place just inside the door. The doorknob vibrated from the music. But I could still hear the plaintive voice of one of my companions, a first-generation Hmong in his late twenties. He said he might as well be castrated for all the attention he got from women, which he summed up by forming a circle with his thumb and forefinger. Most of the women at the party were white, and I wondered if he was referring only to white women. "I don't even know why I came," he said, puffing on a cigarette. The other guy — half Chinese, half Japanese — sucked on his own cigarette and nodded in agreement: "I hear you, dude."

I listened recently to a young Filipino in St. Petersburg, Florida, lament that Asian men were the most unwanted males in the world, and it wasn't even worth trying anymore. What do you mean? I asked. "Just what I said," he told me. "Not worth it . . . they got their minds made up, and there's nothing you can do." Who's got their minds made up? I asked. He looked askance at me with this expression of "C'mon, dude. You know what I'm saying." I could only sigh.

I've exchanged e-mails and letters with a young man from Chicago with a Chinese mother and a white American father. He inherited his mother's Asian features, the luck of the draw unkind. "Why couldn't I have been born a girl? I'd have a better chance," he wrote me. "If it were up to me, I'd reverse the old Chinese practice of killing girl babies and recommend they kill boy babies at birth. It'd be a form of mercy killing."

At one time it was possible to think of these as the lamentations of an isolated few. Outcasts were like the poor in the Bible: they would always be with us. They would be drawn to one another, the magnet of misery. But the coming of the digital age revealed how widespread and deeply ingrained were these feelings of bitterness, of exclusion from the main vein of life, among Asian American men.

Even today, thirty years after my drunken conversations with Cho and a dozen years into the new millennium, you can find Chos everywhere, and they have found one another in blogs and chat rooms and discussion groups, on websites like asianmalerevolutions.com, bigWOWO.com, alllooksame.com, 8asians.com, goldsea.com, asian-nation.org, and dozens more. It would be hard to find a website about Asian America that *doesn't* devote space to the struggles of the Asian male in the Western world.

Phil Yu, a thirty-something Korean American in Los Angeles, is one of the most influential bloggers in Asian America. His website, angryasianman.com, receives 250,000 unique visitors a month. The maligned Asian male has been Yu's running theme since he started the blog in 2001. His first high-profile run-in came three years later against the men's magazine *Details*. The magazine published a spread titled "Gay or Asian?" with a photo of a young spike-haired Asian man in a white V-neck T-shirt, Dolce & Gabbana suede jacket, Evisu jeans, metallic sneakers, and a Louis Vuitton bag across his shoulder. The sunglasses amplify "inscrutable affect," the article said. His "delicate features" are "refreshed by a cup of hot tea," the jacket "keeps the last samurai warm," the T-shirt "showcases sashimi-smooth chest," and we can't forget his "lady-boy fingers . . . perfect for both waxing on and waxing off."

The article was intended as parody, but many Asians found it beyond the pale. "It seriously pulls out every offensive, stereotypical Asian pop culture reference imaginable," wrote Yu, rallying his readers to complain. They complained in sufficient numbers to prompt an apology from the magazine. When Yu's blog isn't angry, it tends to be irreverent, self-satirizing. It features an Angry Reader of the Week, Angry Posts of the Week, and Angry Links of the Week. The logo is a cartoon of a bare-chested Asian martial artist with his kicking foot in your face.

It was on Yu's blog where I first read about Wesley Yang. In a 2011 *New York* magazine cover story, Yang wrote about the "bamboo ceiling" that prevents Asians in corporate America from rising beyond a certain point. Yang is angrier than angry asianman.com. His take on Asian values: "Fuck filial piety. Fuck grade-grubbing. Fuck Ivy League mania. Fuck deference to authority. Fuck humility and hard work. Fuck harmonious relations. Fuck sacrificing for the future. Fuck earnest, striving middle-class servility." These values, Yang says, don't produce men who can compete in the higher echelons of corporate America, which prizes dominant Western-style masculinity. Once upon a time on a continent far, far away, these values served a purpose. In today's Western-dominated world, they handicap.

Yang cites the following statistics: up to 20 percent of Ivy League graduates are Asian, and the Ivies are supposedly "the incubators of the country's leaders." So where are all the Asian leaders? (At Harvard, applicants are sometimes interviewed by alumni. One of the assessed qualities is leadership ability. Until a few decades ago, the leadership box on the interviewer's form was labeled "Manliness.") Asian Americans make up 5 percent of the population but only 0.3 percent of corporate officers, fewer than 1 percent of corporate board members, and 2 percent of college presidents. In Silicon Valley, a third of all software engineers are Asian, but they make up only 6 percent of board members. At the National Institutes of Health, 22 percent of tenure-track scientists are Asian, but they hold only 4.7

percent of director positions. The takeaway from all this? Asians are seen as good worker bees but can't rule the colony.

Yang asks: "What if you missed out on the lessons in masculinity taught in the gyms and locker rooms of America's high schools? What if life has failed to make you a socially dominant alpha male who runs the American boardroom and prevails in the American bedroom? What if no one ever taught you how to greet white people and make them comfortable? What if, despite these deficiencies, you no longer possess an immigrant's dutiful forbearance for a secondary position in the American narrative and want to be a player in the scrimmage of American appetite right now, in the present? How do you undo eighteen years of a Chinese upbringing?"

Jerry "J.T." Tran offers an answer: the first step is to learn how to act like a Western alpha male. Tran is Vietnamese American, an aerospace engineer who left behind turbine engines to become a trainer of Asian men. His area of expertise is romance, but his teachings apply in the workplace, too. He travels the country conducting workshops on how to project oneself in a way that communicates strength, confidence, and daring—the qualities that are generally believed to appeal to Western women and Western corporations. His main course, "The ABCs of Attraction," has drawn sizable audiences on college campuses, including hallowed Ivies such as Harvard, Yale, and the University of Pennsylvania.

One morning in the spring of 2011, Tran strides to the front of a lecture hall at the University of Chicago. He's been invited by the Asian fraternity Lambda Phi Epsilon. The audience, mostly Asian men and women, applauds enthusiastically as Tran scans the crowd with a beaming face. He wears a silver-gray suit over a linen shirt unbuttoned to mid-chest. A burgundy kerchief peeks from a breast pocket. His coiffed hair glistens as he paces the floor and gesticulates. Occasionally he sips from a water bottle, and he takes his time about it. He pauses before and after the sip. The audience hangs on his silence. He doesn't mind. Projected on a large screen is the outline of his talk: *Inner Strength, Outer Confidence, Verbal Attraction*.

If a "short, nerdy Asian guy like myself can be successful with women, anyone can," he says. Then he addresses the male majority in the room. Men: Pay attention here! Listen! This will change your life.

"How do you convey yourself physically?" he asks. "You may be internally confident, you may think the world of yourself, but if you don't actually project it, no one's going to know. People make judgments about you based on how you dress, how you walk, how you appear."

For example, how does an alpha male stand? He answers by showing them: neck extended, spine straight, shoulders back, legs slightly splayed, feet planted slightly wider than shoulders. Do not slouch, do not put your hands in your pockets, do not stare at the floor. *Command the space.*

How does an alpha male walk? Not by shuffling like someone shackled at the ankles. Raise your feet completely off the ground, stride into forward motion with a slight sway in the shoulders. The arms must swing naturally. Tran walks the walk throughout the presentation. Over several hours, he gives advice on areas ranging from the importance of eye contact and clear, confident speech to the proper way to pull off a woman's jeans—in one swift, controlled motion—which he demonstrates to the utter joy of his audience.

The training may seem mechanical, and may smack of Hefneresque male chauvinism. But Tran says the need is great among Asian men. The voice of "the gibbering monkey of fear" in the back of the mind is too loud, its message too persistent, for a lot of men to overcome on their own. The monkey, he tells the crowd, says, "You can't do that. You're going to embarrass yourself. People are going to laugh at you." The monkey gets its material and power from books, television shows, and movies that drive deep the idea of Asians as the runts of the human race. It takes vigilance and sometimes step-by-step training, Tran says, to get that monkey to shut up even for just a night.

7

Tiny Men on the Big Screen

Humbly asking pardon to mention it, I detect in your
eyes slight flame of hostility. Quench it, if you will be so
kind. Friendly cooperations are essential between us.
Wishing you good morning.

— *Charlie Chan, in* The House Without a Key

I know the gibbering monkey well, and feel profoundly slighted
by the Hollywood gods who feed it material. Occasionally I'll
turn off the television or walk out of a theater vowing never
to watch another Hollywood production again. I vow and fail. I
always come back. Occasionally, a new Asian character appears
on the screen who seems to break out of the old molds, like the
character Glenn Rhee in the phenomenally popular television
show *The Walking Dead.* I watch Glenn closely. I root for him.
He does bust through the old stereotypes, but only in moments.
He crosses the magical line and dips his toes in the water of full
manhood, but then he retreats back to being an errand boy. My

moments of feeling hopeful dissolve into disappointment. The monkey gibbers.

At least I did not have to live through the era of Fu Manchu and Charlie Chan, which never bothered to conceal its racism. These two characters — one an embodiment of the Yellow Peril, the other an early incarnation of the Model Minority — were white creations played by white actors in yellowface. By the time I started spending immense blocks of time staring at the television and movie screen, most Asian characters were at least played by actual Asian people, but not all.

I remember watching an old movie on television about the Mongol conqueror Genghis Khan, the title character played by that oh-so-Mongolian actor John Wayne (*The Conqueror*, 1956). The best-known "Asian" martial artist in the history of television, the character Kwai Chang Caine in *Kung Fu*, was played by David Carradine, who to me looked about as Asian as Spiro Agnew. He may have passed muster with white Americans, but to Asians, Kwai Chang Caine was just a white guy in yellowface, and therefore his triumphs did not apply to us. Apparently in the 1970s there were no Asian male actors qualified to play an Asian leading man on television. When actual Asians appeared on-screen at all, they played servants, villains, or geeks — one-dimensional, powerless, sneaky little men with as much sex appeal as a sack of rice.

I have white friends who don't understand why this is such a big deal. "It's just TV," they say. I tell them about an episode of *Bonanza* called "A Lonely Man," about the Chinese house-boy Hop Sing. A diminutive figure compared to the strapping Cartwright men, Hop Sing, with his broken English and braided queue, cracks up the Cartwright brothers by threatening to hit them with his frying pan. Harmless, haranguing Hop Sing. Just like a cranky woman! In this episode he falls in love with a shy young white woman, who is equally drawn to him, but predictably they never consummate their love. At one point a group of town thugs rough up Hop Sing for attempting this transgression, and he cannot defend himself. Ben Cartwright rescues him.

"Hop Sing good as any ranch hand. Ranch hand get married. Why not Hop Sing?" he asks Ben Cartwright, whom he calls Number One Boss. In the end, the woman goes her way, and Hop Sing stands over a cooking pot brokenhearted and weeping. *Bonanza* was one of the most popular shows on television for more than ten years.

Fast-forward three decades and the most recent incarnation of the Weak Little Asian Man is a character named Lloyd Lee on the HBO series *Entourage*. Lloyd takes wimpiness to another level. He is small and effeminate, and openly admits that his best skill is "obsequious ass kissing," applicable for "a lifetime of servitude." He's also gay. He skips around the office and takes ridicule from his Jewish boss, the brash, tough-talking Ari Gold. When Lloyd comes to the office sporting a tight-fitting new suit, Ari tells him he looks like Michelle Kwan in drag. It's a running joke. At one point, Ari implores Lloyd to be a man — or as much of a man as he can possibly be *for god fucking sakes.*

It's just TV, I know. But television, during my first decades in America, was my most important teacher. It taught lessons through repetition. Multiply an image, a message, a hundred, a thousand times, and its persuasive power increases proportionately. The secret to propaganda, and to its more benevolent cousin, advertising, is to repeat something often enough until it passes for truth. Channel it to television screens in every household in the land, replay it relentlessly, and it becomes dogma — so pervasive and routine that it melds with what you thought you already knew. It becomes part of what always was.

By the time I reached the university at age seventeen, I had logged countless hours in front of the television, had absorbed the standard notions of who was vital and inconsequential, who was beautiful and ugly, who was powerful and weak, what it took to get the girl, how to be a man. Men were strong, got things done when others could not. They were brave, protected the weak, took what they wanted. They did what was necessary in service to a greater good. Like Ben Cartwright. Like Lucas McCain in *The Rifleman.* Like Steve McGarrett of *Hawaii Five-O.* This last show really crystallized the racial hierarchy. Even on their own turf of Hawaii, brown and yellow men took their orders from a white guy in a suit. "Book 'em, Danno!" McGarrett barked.

All of it would have been easier to dismiss as fiction if it were not that it seemed to mirror the real world. White men *did* rule the planet. If that part of the grand mythology was true, was the rest of it bound to be true as well? Was I destined to be a houseboy, an order-taker, never in the seat of power or at the forefront of events, never to have the wherewithal to impose my will upon circumstances, upon rivals? Never to get the girl? Was my fate to be always in the background, merely a witness to power in action, an assistant to The Man?

Movies offered no more hope. I was eight when Sean Connery first mesmerized me in the James Bond film *You Only Live Twice.* My brothers and I thoroughly enjoyed it. But it was a different experience watching it as a teenager at the university. Fridays at McAlister Hall was movie night, and our regular fare was the raunchiest pornography we could get our hands on. The movies featured the usual white and black studs with white women and the occasional Asian woman. Asian men never appeared. They were not part of the sexual equation.

I remember holding a Heineken and thinking to myself, *So what.* So fucking what. *It's just porn.* But it was hollow consolation. It's possible the young women in our Friday night group did not notice or care, but I cared. It mattered to me. It mattered in a secret subterranean way, as if my gut knew instinctively that respect grew from below as well as above. Thugs, fighters, rogues, outlaws, and yes, porn stars—dwellers of the underbelly—push masculine esteem through cracks at ground level, and males, young and old, schooled and unschooled, secretly admire them. At least the idea of them. And somehow these role models from the nether regions add a dimension to our common understanding of what makes a man.

When an Asian man finally did appear in one of our Friday night movies, he brought little solace. Occasionally we ditched the porn and watched action movies, and I remember in the middle of *You Only Live Twice* feeling embarrassed by the character Tiger Tanaka. In Ian Fleming's book, Tanaka, the Japanese secret service chief, is a formidable man on more than equal terms with Bond. In the movie, Tanaka is no tiger. More a housecat. Domesticated, de-clawed. When the two first meet, they stand face-to-face, Tanaka markedly shorter than the towering Bond. Bond asks for the mission password, and Tanaka utters submissively, "I love you." Tanaka officially outranks him, and the setting is his home turf of Japan, but Bond's superior manhood is instantly established. Bond goes on to manhandle armies of Ninjas and Japanese fighters sent to foil him. At one point, Bond and Tanaka sit in a bath having their backs scrubbed by young Japanese maidens. The women show coy curiosity about Bond's hairy chest.

"Japan men all have beautiful bare skin," Tanaka says.

Bond, grinning winsomely, replies, "Japanese proverb say, bird never make nest in bare tree." Giggling all around. Superior manhood asserted. Conquest complete.

The tame Tanaka isn't an obvious caricature like the buck-toothed Mr. Yunioshi in *Breakfast at Tiffany's*, whom you could laugh at and forget. Tanaka is just plausible enough that you could absorb his implied inferiority and not think twice about it. That's how it works. Messages hidden in the thickets of a story are the ones that burrow deepest because most of us don't realize that any burrowing is going on at all. I could almost feel my dorm buddies at McAlister Hall absorbing the message that night. Sitting there buzzed after a couple of beers, I recalled the two young women, Lisa and Rosemary, on the Grand Concourse five years earlier. They had mistaken me for Japanese, and I didn't mind one bit. Tonight I would have minded.

Fast-forward a couple of decades to *The Last Samurai*. The Tom Cruise character, Nathan Algren, kills a Japanese warrior and wins the love of the dead warrior's beautiful wife, and in the end outlasts the last real samurai. The movie pays homage to Japanese culture, but the outline of the story is a tried-and-true formula: a Western man goes east, conquers the men, takes the women, and brings a resolution to some Eastern quandary. The movie ends with Algren returning to claim the lovely Taka, played by Japanese actress Koyuki. I liked this movie, as I do many of the movies I mention here. I just wish the formula could work in reverse, too. Imagine a movie in which a Japanese man comes to America, outfights the Army Rangers, kills one and beds his wife, saves the world, and settles down with his new bride in a picket-fenced house in the suburbs. I doubt this movie will be coming to a theater near me anytime soon.

White men claiming Asian women happens routinely—in movies as in real life, in settings foreign and familiar. They include the films *Careless Love, Sideways, The Quiet American, Silk, Cypher, Snow Falling on Cedars, Come See the Paradise, Chinese Box, Red Corner, Double Happiness, M. Butterfly, Indochine, Tai-Pan, When Heaven and Earth Change Places, Tomorrow Never Dies, You Only Live Twice, The Barbarian and the Geisha, Sayonara, The World of Suzie Wong,* the television movies *Shogun* and *Marco Polo* (1982 and 2007 versions), the opera *Madama Butterfly,* and many others. With more to come. A screen version of *Miss Saigon,* about the relationship between an American GI and a Vietnamese bar girl, is said to be in the works. Western men claiming Eastern women is an accepted cultural trope. We could debate whether movies reflect or create the reality. It's clear to me they do both.

Hollywood has not minded portraying Asian male power as long as the Asian male conforms to known clichés—sage, brainiac, martial artist—and keeps his genitalia in his pants. I can't recall an American movie that remotely approaches the steamy French film *The Lover,* about an affair between a French teenager and a wealthy young Chinese businessman (based on the real experiences of writer Marguerite Duras). The male lead, played by Hong Kong actor Tony Leung, exhibits raw sexuality and no kung fu whatsoever. But, as Sheridan Prasso observes in *The Asian Mystique,* Hollywood studios seem loath to let Asian men get too romantic on the big screen, particularly with non-Asian actresses.

In the 2000 movie *Romeo Must Die,* Hong Kong action

star Jet Li plays the Romeo character opposite the late singer-actress Aaliyah Haughton, who plays the Juliet. Aaliyah, young and beautiful, appears ripe for romance. I find myself eagerly awaiting consummation, like Romeos and Juliets do. In the end, after Li defeats the bad guys, Aaliyah gives him a hug. A kiss scene was filmed but edited out just before the movie's release. The following year, Li played opposite Bridget Fonda in *Kiss of the Dragon*. Again, chemistry, tension. In the end, Fonda gives Li a peck on the back of his hand. In *The Replacement Killers*, another Hong Kong actor, Chow Yun-fat, plays opposite Mira Sorvino. Chow plays a credible alpha, and there are teasing moments of possibility, but nothing romantic pans out. The same with Jackie Chan and Jennifer Love Hewitt in *The Tuxedo*. The I-can't-stand-you banter that usually leads to a locking of lips goes nowhere. The same with Chan and almost all his American female co-stars. The studios won't let him close the deal.

My nephew Caleb has a theory about this. He was a fan of the HBO series *Sex and the City*. Something like 2 million Asians live in the New York metropolitan area, but Asians hardly appeared in the show at all — symbolic annihilation at its best. That's a real sociological term, by the way: *symbolic annihilation*. It feels the way it sounds, and applies here. I tried watching *Sex and the City* with Caleb and I felt annihilated. Men like Caleb and me did not exist in the universe occupied by the beautiful and quirky Carrie, Charlotte, Miranda, and Samantha. They dated all kinds of men — a Russian, a Latino, a couple of African Americans, several Jews, but no Asians. Asian men did not figure in as men. As props, yes. As spectators. Watching those few episodes, I felt like a spectator gazing through the beveled window of an exclusive club.

"None of the women on the show would be attracted to an

Asian," Caleb said to me. He's twenty-one, half Filipino, half white, a college student and a TV and film aficionado. He once described himself to me as "an ashamed Asian," and he said it like it was an official category. "If one of the women did have a thing with an Asian guy, the people watching the show would be . . . would be . . ."

As Caleb searched for the right word, a few options popped into my head: *Put off? Pissed off? Turned off? Grossed out? Emotionally traumatized? Permanently scarred? Haunted by visions of the impending Apocalypse?*

"Alienated," he said finally.

"Alienated?" I said.

"Yeah. They wouldn't be able to relate to it, and they'd be turned off by the idea. That's why the show won't do it — ever. That's why the movie versions will never do it. They don't want to turn off the audience."

Does the same reason apply even when the female lead is Asian? In the 2010 Lifetime movie *Marry Me,* the protagonist, played by Chinese American actress Lucy Liu, is courted by three dashing Prince Charmings, all white. If the female lead were black, would the show's producers have dared exclude black men from the cast of courters? If white, would they have excluded white men? I doubt it. The producers would have included at least one Prince who was similar in racial appearance to the leading lady.

The annihilation happens routinely, seemingly without any thought to reflecting actual demographics. The 2007 movie *30 Days of Night,* starring Josh Hartnett, is set in the town of Barrow, Alaska, whose real population is three-quarters Native. (Alaska Natives have racial and cultural ties to indigenous peoples of Siberia.) But none of the main characters are Native, and

Native-looking people barely appear in town scenes. You would never guess from watching the movie that Barrow is essentially a Native village. I was struck by the discrepancy only because I have spent time in Barrow and traveled all over Alaska, and I know how Alaskan Native villages look and feel.

I've also lived and studied in Hawaii, and I know that 50 percent of the population identify as Asian or Pacific Islander, and one quarter as white. But after watching the 2011 movie *The Descendants*, starring George Clooney, you would think Hawaii's population was made up entirely of whites, with just a few Hawaiians around strumming ukuleles and occasionally serving drinks. It was one of my favorite movies of that year. Couldn't the producers have given at least one speaking role to an Asian or Pacific Islander man for the sake of accuracy, or even just to make the movie more plausible for people who actually live or spend time in the state?

The Hollywood gods take it one step farther: it's perfectly acceptable to exclude Asians, even to erase them altogether, in movies based on true stories about real Asian men. Recently I watched the 2009 movie *Hachi: A Dog's Tale*, about the undying loyalty of a dog to his master, a university professor. I knew the real story because I'd owned an Akita — same as the real Hachi — and was familiar with the breed's history. The Akita is a distinctly Japanese dog, and part of the real story was the incredible loyalty that characterizes the breed. The real-life professor was Hidesaburo Ueno of the University of Tokyo, who died in 1925. The story of his famous Akita is memorialized with a bronze statue in Japan. In *Hachi*, the dog's breed is the same, but the setting is New England and the professor is Parker Wilson, played by Richard Gere. All right, so the movie was an

American adaptation. Adaptations of foreign films are becoming more common these days.

But the studios don't mind doing the same even when it involves true stories of Asian *American* men. The film *21*, which topped the charts for a month in the summer of 2008, was based on the nonfiction book *Bringing Down the House,* about a crack blackjack team from MIT that takes Vegas casinos for millions in winnings. The main character is based on a true-life ace, Jeff Ma, a Chinese *American.* Most of the real MIT team was made up of Asian *Americans.* The man who led the team was a real Asian *American* professor named John Chang. But in the movie, Ma's character is played by white British actor Jim Sturgess, the Chang character is played by Kevin Spacey, and the team has only two Asians, both minor characters.

"Believe me, I would have loved to cast Asians in the lead roles," wrote *21* producer Dana Brunetti on an entertainment blog. "But the truth is, we didn't have access to any bankable Asian American actors that we wanted."

As I write this, Tom Cruise is in pre-production to play the lead role in *Edge of Tomorrow,* a Warner Bros. adaptation of the Japanese novel *All You Need Is Kill* by Hiroshi Sakurazaka. Cruise's character, Bill Cage, replaces the Japanese character Keiji Kiriya. Warner Bros. is also in pre-production on a movie adaptation of a Japanese graphic novel, *Akira,* by Katsuhiro Otomo. The lead character — Shotaro Kaneda in the novel — will be played by blond, green-eyed Garrett Hedlund.

Why not Japanese or Japanese American actors for these parts? Insert Dana Brunetti's statement here. It's an all-weather excuse. Hollywood producers won't heed the concerns of underrepresented people until not heeding them hurts profits. The studios will claim there are no bankable Asian male actors and will exclude them from star-making roles, making it even harder for an Asian male actor to become bankable. As in *Hachi* and *21* and a bajillion other examples, the producers will include a token Asian presence and call it good.

Meanwhile, they'll hire Asian males to play the likes of Mr. Miyagi, the sexless sage of the *Karate Kid* movies, or the slimy and flamboyant Mr. Chow of the *Hangover* movies. They'll get them to play incidental roles like Mike Yanagita in *Fargo*. Yanagita is a pudgy, bespectacled Japanese American engineer who, after failing to seduce police chief Marge Gunderson, begins to blubber "I've been so lonely" into his drink. Marge, the seducee, ends up comforting him. *Here, here, little guy, it's okay.* Yanagita is a passing character, but I remember him vividly, as I remember lots of other fleeting appearances of Asian men.

Examples off the top: the apron-wearing Mr. Lee in both versions of *True Grit* (foreigner). The crewmember Ravel in *Prometheus* (nonessential). The otherworldly Mr. Wu in the remake of *The Day the Earth Stood Still* (incidental, alien). The human Gumby Yen in *Oceans Eleven, Twelve,* and *Thirteen* (shrimp). Bruce the MIT grad in *Get Smart* (geek). The vertically challenged Yin Yang in *The Expendables* (comic relief, shrimp). The servant Yuan in *Just Married* (houseboy). Loyal friend Ronnie in *Disturbia* (sidekick). The student Choi in *21* (incidental). The unnamed airport Asians in *Up in the Air* (props). In this last one, the George Clooney character notes a group of businessmen going through a security checkpoint. Asians. They pack light, travel efficiently, and have a thing for slip-on shoes, he says, with evident admiration. When a companion calls the statement racist, Clooney explains that he stereotypes because it's faster.

• • •

A recent conversation with a (white) friend:

FRIEND: It'll happen. It's coming. There'll be an Asian
   leading man who'll break through. Some Chinese or
   Korean dude with star quality, crossover appeal.
ME: There haven't been any Asian actors with star
   quality?
FRIEND: Not really. I mean, there're the kung fu
   guys—the Bruce Lees, the Jackie Chans—but they're
   not it. They're niche. I'm talking crossover. General
   audiences. That guy's coming.
ME: In my lifetime, you think?
FRIEND: Could. Be patient.
ME: Be patient, huh? Easy for you to say.
FRIEND: I know.

Progress has been slow for those of us waiting. The most
promising signs have come, now that I think about it, from
where you'd expect them to originate—the East. The 2000 movie
Crouching Tiger, Hidden Dragon, starring a Hong Kong cast, won
four Oscars and became the highest-grossing foreign-language
film in U.S. history. Four years later, China's *House of Flying Daggers* did not rake in the same profits but won critical acclaim and
further inspired American filmmakers to emulate balletic high-
wire martial arts. Then in 2008, *Slumdog Millionaire,* based on
an Indian novel, set and filmed in India with an Indian cast, won
eight Oscars, including Best Picture. In 2012, one of the most
critically acclaimed films of the year was *Life of Pi,* about a teenage boy from India, and directed by Ang Lee, who was born in
Taiwan. The movie was an Indian-British-American production.
   Hollywood will surely follow the money, but it still lags. Prog-

ress has come in increments, tiny encouragements, followed by long periods of nothing. Asian actors who seem to signal a new era emerge every so often — Tony Leung, Russell Wong, Jason Scott Lee, Rick Yune — but the star-making lead roles, those breakthrough performances that change the cinematic landscape, have not materialized.

Asian actors *have* broken farther out of the martial arts box. John Cho, the Asian American actor of the moment, gives credible performances in *Star Trek* and *Total Recall.* As Kato in *The Green Hornet,* he's back in the box. While on more equal footing with the Hornet than in the 1960s television series, Kato is still only a sidekick. Cho's best-known role, in which he co-stars with the Indian American actor Kal Penn, is as junior banker Harold Lee in the *Harold & Kumar* stoner comedies, in which he does eventually get the girl. Genteel audiences dismiss the *Harold* movies as, well, stoner comedies, but they have brought new cachet to Asian characters, at least in the murky depths of adolescent hearts and minds.

Japanese actor Ken Watanabe has played complex masculine characters in movies such as *Inception, Letters from Iwo Jima, Memoirs of a Geisha,* and *The Last Samurai.* This last one is a Tom Cruise vehicle from beginning to end, but Watanabe's portrayal of the samurai poet-warrior Katsumoto earned him an Oscar nomination and high praise from critics. For most of the film, Katsumoto is Nathan Algren's superior — until the end, when fighting side by side in the final battle, Katsumoto dies and Algren survives.

Bravo to Clint Eastwood for directing *Letters from Iwo Jima,* which tells the story of that epic World War II battle from the Japanese point of view. Eastwood also directed and starred in the 2008 film *Gran Torino,* which centers on the Eastwood character's cranky relationship with an immigrant Hmong family next door. We glimpse the challenges faced by Asian immigrants today. But the Asian males in the story are invariably thugs, geeks, incomprehensible immigrants, or adolescent boys. I keep waiting for a strong, exemplary Asian man to show up, but it doesn't happen. Eastwood plays the only powerful male character you can root for, although the ending offers hope. He bequeaths his prized Grand Torino to the Hmong neighbor boy for whom he's developed a fatherly affection, signaling a passing of the torch: perhaps the boy will grow up to be the strong and sympathetic Asian man missing throughout the movie.

On television, one of the brightest signs, as mentioned earlier, is the character Glenn Rhee in *The Walking Dead*, about a ragtag group fighting to survive through a zombie apocalypse. Glenn, played by Korean American actor Steven Yeun, is in his twenties and delivered pizzas before the world went to hell. He's smart, sensitive, and feisty; he's fleet-footed, resourceful, and exceedingly brave. But what Asian pop culture websites have been buzzing about most is that Glenn hooks up with a tall, voluptuous Irish American farmer's daughter named Maggie Greene, who's pretty feisty herself. In season two, Maggie shocks Glenn — and much of the Asian (the series is quite popular in China) and Asian American viewing audience — by announcing to him in the middle of a supply run: "I'll have sex with you." Where? *Here.* When? *Now.*

"Really?" Glenn says incredulously. "Why?"

Maggie, a take-charge kind of woman, says, "You're asking questions?" and begins stripping. Only after she's taken off her shirt does Glenn start doing the same. The scene fades out. Later, my white friend who counseled patience told me, "There you go. An Asian guy gets it on with a white girl on television.

They're banging each other the rest of the season. They'll probably be banging the rest of the next season, too. Feel better now?"

Sort of, I said. It's definitely progress.

But let's really look at the character. Although he breaks new ground, Glenn also occupies a lot of familiar territory. He's stereotypically thin, on the verge of wispy. He's got a high voice. He takes verbal abuse without fighting back or even showing anger. Others call him "little man" and "short round." In the first season, one of the rougher main characters, in a backhanded compliment, tells him, "You've got balls for a Chinaman." Glenn responds tepidly, as you'd expect an Asian to do. "I'm Korean," he says. It becomes a running joke.

On two other occasions, Glenn is captured by outsiders and rescued by his group, the kind of role typically given to female characters. During the second abduction, he's beaten up and his girlfriend is assaulted. Glenn is tormented by the fact that he could not protect her. Later, Glenn gets in a fight with one of his abductors, the rogue character Merle Dixon, who winds up on the ground on top of Glenn. Guess who comes to Glenn's rescue? His girlfriend Maggie. She applies a killer chokehold on Merle and peels him off her fallen and bleeding boyfriend. And this is actually the second time in the episode that Maggie saves Glenn from Merle. Being rescued by a wife or girlfriend would be unimaginable for any of the alpha males of the group, the true leaders, the manly men: Rick and Shane and Daryl.

Though Glenn is stout of heart, the other men affectionately refer to him as a boy. He fetches peaches for them. He runs errands for them. This is what makes Glenn so appealing to so many viewers. Like the Hmong teenager in *Gran Torino*, he shows the promise of the strong and capable man he might grow up to be. He's likeable in the way that an earnest and pure-hearted boy can be likable. But that's exactly what he is: a boy among men. The show's producers let him get the girl, let him bash an occasional zombie, like everyone else on the show, but they aren't quite ready to make him a man among men.

Ultimately, I believe the tectonic shift I long for will be brought on by people behind the scenes, sometimes very far behind the scenes, in the boardrooms of far-flung and sometimes unexpected enterprises. For instance, in May 2012, the Dalian Wanda Group, the Chinese real estate company owned by billionaire Wang Jianlin, purchased AMC, the second-largest film exhibitor chain — with five thousand theaters — in the United States. The speculation, and the hope among many of us, is that transactions like this will eventually pave the way for more Chinese films to reach American movie screens.

And it's not just Chinese buying into the American scene. The reverse is also happening. DreamWorks Animation plans to build a production studio in Shanghai. The owners of 20th Century Fox bought a major share of the China-based film distributor Bona Film Group. Walt Disney and Marvel Studios teamed with a Beijing company, DMG Entertainment, to produce *Iron Man 3*. With all this cross-pollination going on — and it's just the beginning — you would think the Asian presence in films would *have* to expand to roles beyond stereotypes. Right?

So I vow and fail. I keep watching movies. I keep flipping on the television. I smile inwardly at the small advances, and I swear under my breath at the embarrassments. I hold my breath every time an Asian man appears in a scene. It's me up there, Hollywood. Movies and television shows are magic mirrors, and we watch for the reflections that resemble us. They show us who we are. They give us the silhouettes of identity so difficult to envision on our own.

When Mike Yanagita blubbers into his drink, it's me blubbering. I wish it weren't the case. And I suspect it wouldn't be the case if I had been exposed during my growing-up years to a multitude of like-me characters: heroes and villains, winners and losers, lovers and fighters, geniuses and morons and every variety of ordinary person, and mixtures of all of these. If movies and television had shown me a million renditions of the Asian face, then the occasional blubbering idiot would not have so much riding on him. He would be just another character.

I have another nephew, Kai. He is thirteen, half Laotian, half Filipino, and all-American in his upbringing. He reminds me of me at thirteen. For a few weeks every summer, he comes to stay at my house in Seattle. One night we drove to a video warehouse to rent movies, and for twenty minutes I watched him walk the aisles, gazing at titles, studying covers, and re-walking the same aisles.

"What you looking for, buddy?" I asked.

He shrugged his shoulders. "I don't know."

Finally, I saw him settle in a corner of the store he'd perused at least once already. It was the martial arts corner, starring the usual cast of characters and a few new ones who looked just like the old ones. I walked over to him, pretending to be shopping, but was furtively glancing at the expression on his face as he scanned the covers. Placid resignation. *This is all there is.* I knew what he'd been looking for even if he didn't.

8

Its Color Was Its Size

The dirt road and I, we resemble each other

— *Meifu Wang*

After college and before the working world, I spent a few years on my own walkabout, wandering the country in an old mauve Volkswagen Rabbit that veered left. I had replaced the passenger seat with a jury-rigged cot, made of two-by-fours and a camping pad, that extended into the back seat. There were exactly two books in the driver's seat pocket: a leather Bible from a girl concerned for my soul, and a junky paperback copy of *On the Road*, which I read at night until I fell asleep or until the overhead light began to flicker. When I got to the last page, I would start over. Some nights I felt I could have died in the pages of Kerouac's words, satisfied that I had glimpsed life, even for just a moment, through the lens of an original vision. The guy tapped into the vein. He gave words to the unnamed longing in my gut. Longing was what it was all about in those years. Whatever I set my eyes on, I longed for. What I could not see, I longed for even more.

I went through a period of intense religiosity in my late teens and early twenties, embracing an evangelical Christian faith that made sense of things for a while, that gave me an ultimate destination to long for. But the more evangelicals I met, the less appealing heaven got. I hoped that God would be more interesting than his human ambassadors. While I did meet Christians who intrigued and challenged me, even inspired me, they were few and far between. In any case, by the time of my walkabout, I was reading Tolstoy and Thomas Wolfe and Jack Kerouac more than Matthew, Mark, Luke, and John. I kept the Bible close as much for nostalgia as for hope.

One summer in the village of Cordova, Alaska, as I was living out of my Rabbit, I met a young woman who became my secret girlfriend. People around town saw us together, but we did not act the part of a couple in public. We could easily have been two co-workers, which we also happened to be. Gwen was slightly shorter than me, with wide green eyes and light brown hair that turned blond in certain shades of light. Most people never glimpsed her hair because she always wore a hooded raincoat and a hooded sweatshirt under that. It rained a lot that summer. I, on the other hand, saw a lot of her hair. It was one of my favorite things about her. She had a soft gamine face. She was smart and affable and quick to laugh. Too quick sometimes. She would start laughing way before the punch line, so eager was she to like and be liked. It was something we had in common.

We spent lots of time in my Rabbit. One of our main activities was oral sex, mostly her giving it to me as I sat in the driver's seat. I guess that was the secret part. Gwen loved doing it, and she did more over that summer to assuage my adolescent anxiety about sex than anyone to that point.

We had other things in common: we came from large Catholic families broken by divorce, we liked books, we both had recently finished college with no inkling what to do next. She had ended up in Cordova by way of Indiana, driving in a caravan with three other friends. They had traveled up the Alaska-Canadian Highway just days ahead of me.

Cordova was a fishing village on the southern tip of the Schwan Glacier in Prince William Sound. From the air, a view I sighted in a subsequent trip, the town appeared negligible, accidental, like a cluster of barnacles clinging to a rocky coastline. The town slumbered most of the year, but the fish canneries there drew hordes of itinerant wage seekers in the summer months: illegals and vagrants and fugitives of every persuasion, exiles of all varieties, dropouts, freaks, eccentrics, and survivalists; grizzled frontiersmen who trapped beaver in winter, drank whiskey in spring, and staggered into town in summer, pickled and broke, to sling halibut for $8.60 an hour. The college students from the Lower Forty-eight usually came in groups. They were (mostly) white, wide-eyed, fresh-scrubbed cherubs looking for adventure, a summer reprieve, something beyond life in the cul-de-sac. Gwen and I met on the Slime Line, which wasn't actually a line but a circular metal table where we gutted pinks for twelve to twenty hours a day. Our gentle acquaintanceship began in a mist of blood and viscera while holding sharp knives.

One other thing we had in common was that we were relatively inexperienced in matters of sex. Fear and religion had held me back from total abandon during my few sexual experiences in high school and college. Even when I did somehow win over a girl, it was sinful to plunge into sex before marriage. There is no combination like paralyzing fear and browbeating religion to quash friskiness. I was a friskless young man.

But even before I met Gwen, I had convinced myself that oral sex was not really sex and therefore was okay in the eyes of the Lord. Gwen grew up sheltered and shy. She had discovered the joys of fellatio the summer before and had cultivated the practice on three fleeting boyfriends. The first time we did it in my Rabbit, she called my penis "sweet," which I didn't know how to take. Was it sweet as in "Sweet Jesus!" or sweet as in "Aw, isn't that *sweet?*"

The latter, she said.

"Sweet is good!" she added after studying my face.

"Yeah, I know," I said. "Sweet's better than sour."

"Sweet's *very* good."

Like every young man in the Western Hemisphere, I had put a ruler to my penis, hoping to get a read on my place in the world order. And for the longest time there was a disconnect between my ruler and my own eyes. My ruler told me I was average, if indeed average was between five and seven inches in length and five around, as is almost universally agreed upon. I fell exactly in the middle of average. Yet whenever I looked down there, especially unaroused, my penis often seemed undersized to me. My eyes found a different measure. Why was that?

I now understand this to be common among males, part of the chromosomal package. Maybe there's some hardwiring in the XY psyche that predisposes us to insecurity in this particular area. We all want an oak tree between our legs, a frightening weapon that elicits gasps, inspires fear, a club, a bludgeon, a battering ram to smash open the gates. We fantasize splitting the world with our thrusts. Even men with true oaks wish their wood bigger. The burgeoning of Internet porn has given anyone with a smartphone easy access to endless photo galleries of monster cocks. Monster cocks to infinity. In researching this chapter, I glimpsed enough man meat to last several lifetimes and to understand why, these days, average isn't enough. Not in this realm of caveman clubs and two-pound truncheons. Average is the new tiny.

As a young Asian man, I had to deal with an added dimension to the insecurity. I understand now that my self-appraisal was at least partly colored by the cultural indoctrination I had absorbed much of my life. We all, to some degree, absorb the mythologies around us, our vision refracted by the prisms of our particular time and place.

In his last novel, *Just Above My Head*, James Baldwin writes, regarding the mystique of the black male penis, that "it was more a matter of its color than its size . . . its color *was* its size." Baldwin was saying that black, seen through the lens of our culturally constructed expectations, was often enough to create the mystique of virility. Of size. Sometimes you see what you expect to see, or what you *want* to see. Its color was its size. Perhaps the same dynamic applied to yellow men in reverse: its color was its *lack of* size.

Perhaps Westerners see in the Asian man — and by extension his penis — what they expect to see, what in some ways they need to see. Let's be honest: many Western men derive enormous comfort and a good measure of delight in the myth of the small Asian penis. Racial *Schadenfreude* is why the myth endures, and why it will continue to do so. Too many non-Asian men get an immeasurable ego lift from it.

Never mind that no comprehensive science has conclusively verified the myth. Never mind that Google can provide a fair number of visual examples of Asian oaks. Never mind that the myth deeply hurts young Asian men in the West attempting to forge a sexual identity in the midst of overcoming a host of other demeaning perceptions. That it hurts them in a similar way that the myth of an undersized intellect hurts young black girls and boys. This latter myth tells young blacks that they're intellectually inferior. The former myth tells young Asian men that they're sexually inadequate. Who's to say which is more wounding to the spirit in a young person seeking reassurance of worth and belonging, and trying to gauge how far he can expect to go in the world? I can say from experience, corroborated by a vast body of literature, that from puberty to marriage, almost nothing occupies the male brain more than sex, courtship, and love — and sex above all.

As an adolescent in America, I would have preferred virility to intellectual vigor. A perception of virility would have allowed me entry, I believed, into the game I most wanted to play. To fail there meant utter loss. To be excluded from this realm felt like exclusion from life itself, banishment from the essential core of things, from the only thing that truly mattered. The main event.

Instilling the message in a young man that he has a small (read: inadequate) cock cuts him down to size, informs him of his failure, his essential inadequacy, which can only mean that the one relaying the message must be more adequate than he. Spreading the rumor of an entire race of small-cocked men gives the rumormonger something to stand on to make him feel bigger. And it must mean that the monger's race is more masculine, more worthy of the affections of women, more deserving of respect and admiration from other men.

So when the previously unknown Jeremy Lin, then of the New York Knicks, lit up the scoreboards for a phenomenal three

weeks in the winter of 2012, Fox Sports columnist Jason Whitlock made sure to cut him down to size. Lin is an American-
born Taiwanese, a yellow man amid a throng of towering black
and white men. And Lin sometimes looks even shorter than his
actual six feet three inches because, perhaps, his color is his size.
He is thinly but adequately muscled, quick and agile, and absolutely fearless. The sight of him burning up the court and running circles around the NBA's most celebrated athletes stunned
everyone who witnessed it.

This wasn't Ping-Pong. This was a high-profile, high-testosterone game involving the tallest and most athletic male speci-
mens on the planet. Overnight, Lin became a hero in the yellow
quarter of the globe. Obscure little villages in China celebrated
him. Then, just after his most dazzling performance, in which
he scorched Kobe Bryant and the Los Angeles Lakers, the columnist Whitlock, who is black, tweeted to his followers, "Some
lucky lady in NYC is gonna feel a couple inches of pain tonight."
Translation: This Asian dude might have outplayed the brothers
tonight but he'll still come up short under the sheets. Take that,
uppity yellow man.

Within hours, the tweet went viral. For the next several
weeks, Lin's sexual competence became a running theme in the
news coverage. Here's the lead of one article in the online *International Business Times*: "Rumors about Jeremy Lin's manhood
size, which include speculation about how 'big' he is and his
ability or inability to please a woman, have swirled online and
in private conversation ever since he broke onto the scene as
the New York Knicks' new 6-foot-3 Taiwanese-American point
guard."

The guy was just trying to play good basketball, and suddenly
the public discussion turns to his cock. And not just his alone,

but that of everyone who shares the same racial uniform. The entire Asian male franchise fell under scrutiny.

Imagine a young Asian American man reveling in Lin's success, in his exertion of will and skill in a forum where Asian men have been all but invisible. Imagine this young man reading Whitlock's tweet. The message is that no matter what he achieves, an Asian man must still bear the stigma of being less well endowed than other men. The aura of smallness follows him wherever he goes. It's a pall that envelops him. It's a subplot in the story of every Asian man in the West.

It's why this chapter was required: you cannot deconstruct the experience of the Asian male without discussing the myth of his small penis any more than you can examine the black male experience without broaching the myth of his large one. Both mythologies stem from a broader premise that positions white males as the "just-right middle," the ideal: not too big, not too small. From this view, black men, as embodied by their mythically large members, are inclined toward one extreme of manhood (primitive, beastly, dangerous), and Asian men, symbolized by their mythically smaller parts, are inclined toward the other (effeminate, passive, weak). It all works out well for white men (balanced, proportionate, beneficent), who, as members of the most evolved strain of *Homo sapiens,* occupy the hallowed center, where all men should aspire to be.

Its color was its size.

When I was fourteen, living with my family in the Bronx, I had a best friend named Vincent. We were in the same class at JHS 79. He was tall and lanky, with an air of sweetness and innocence that was not entirely phony. He was a pale-skinned, blue-eyed Jew who grew up in Argentina and who moved with his mother and sister to the Bronx the same year I did. We prowled the neighborhood together, a couple of peace-loving outsiders trying hard to blend in. We did the typical teenage things — went to movies, ogled girls, talked sports, traded pornography, and compared masturbation techniques. We spent a lot of time talking of everything to do with girls. They were still a largely undiscovered country. We fantasized about the ecstasies we would find there.

Vincent and I used to play a game in which we'd spot an attractive woman in a store or some other public place, get as close to her as we could without being obvious, and then — right there in the aisle or wherever — we'd secretly show each other the bulges in our pants. We were fourteen. We were squirrelly and obsessed and got erections at will. We always maintained a respectful distance from our targets. One woman in a music store, a Puerto Rican with a hellacious figure, caught on to our game and winked at Vincent. She told him something in Spanish as she swiveled away. Vincent preened.

During sleepovers in Vincent's apartment, we'd compare equipment. We never touched each other's genitalia, but we did share specs, like a couple of boys comparing model cars. Once we held a contest to see who could ejaculate farthest, and if I remember right, we both approached nine feet (or was it six feet?) but I bested him by a head. That I do remember. I raised my fists in the air like Muhammad Ali over Sonny Liston.

Vincent was proud of his penis. It was thick and longish even when flaccid, although it wasn't as long as he claimed. He added a half inch or so. As a matter of general policy, you should never believe anything a male says about his penis, and this includes anything I say about mine. That's because the matter isn't only about inches. It isn't even *mainly* about inches. It's about men measuring their value as sexual beings. With so much at stake, it's hard to keep the ruler still. Anyway, Vincent's penis, when aroused, stood at a slight upward angle, the base somewhat thicker and tapering toward a nice big head. It did not grow all that much from its flaccid state. It filled out, certainly, and elongated a little, but what you saw soft was mostly what you got hard.

"God, it's so beautiful!" he used to say, clasping it like a sword. "Don't you think it's beautiful?"

"Uh-huh," I'd say.

Luckily for me in our comparison game, my penis grew substantially when aroused, although, like my later Rabbit, it veered slightly left. When Vincent and I held our erections side by side, I was always pleasantly surprised to see the marked *lack* of difference. His was slightly longer and straighter, mine, we both agreed, was slightly darker and more mysterious. Neither was the hands-down leading man, although if forced to choose, I would give the nod to Vincent. So would he. We men, who will always be partly boys, measure differences by fractions of an inch, and Vincent gloated about his slight (maybe three quarters of an inch) advantage over me.

My black protector-friend at school, Joe Webb, claimed to have an eight-inch cock but also once confessed sheepishly that he climaxed quickly after entering a girl. "I bet you chinks and white boys with your pinky dicks last longer," he told Vincent and me. The two of us were virgins and could only nod in awe and try to imagine the orgiastic world that our friend claimed to live in. Those black guys sure had it good, we thought.

It would not be until college that I began learning the downside, the shadow reality, of the big black penis myth. A handsome coal-black African American guy named Raymond, whom I met my freshman year, confessed to me over vodka that he was "very average." He said women were sometimes disappointed when the big unveiling took place. "They'd have this expression, like, 'Oh.' And you could tell they're thinking 'Where's the rest of it?'" he said. But he did not seem to carry any shame about it. At one point he told me he liked being average.

Why is that? I asked him.

"Some brothers get down on themselves for their big dicks," he said. "A big dick must mean a small brain, right?"

A black man's large penis pointed to the possibility of truth in other stereotypical notions, "and that's a path a lot of brothers don't want to go down," Raymond said. It all sounded so adolescent. Yet Raymond and I understood that we lived in a world where puerile thinking was an undercurrent, always there, always just a moment away from surfacing.

Raymond and I asked each other questions we couldn't answer: Would embracing one stereotype require you to embrace all? Would shunning one require shunning all? What exactly is the honorable way to behave toward a stereotype, a mythology that we believe contains a nugget of truth? We doused these questions in Smirnoff over many months, reveling in the camaraderie even as we resigned ourselves to vexing conundrums. What do we do about this race thing? Is there any way around it? We often reached a point in our Smirnoff sessions when we would sigh and fall silent. Apprehending it all felt impossible. We could not even think about transcending it. "What a fucking cauldron," Raymond used to say.

For a while, during my wandering twenties, I dated a pretty red-haired woman named Jill. On our first night out together, after an evening walk around Green Lake in Seattle, we found a park bench and started making out under moonlit tree branches. In the middle of our kissing, without any preceding conversation, she said, "You know, the whole size thing . . . we girls don't care about it as much as guys do. It'll be fine. I'll like yours."

We had not gone beyond kissing — okay, maybe a little beyond kissing — and already she was reassuring me about the size of my penis. Did she say that to all her dates? Or did she say it to me because of who and what I was? Did I somehow signal an apprehension about it? We're all so inexpert about ourselves.

I didn't know how to respond to her comment. I remember feeling pumped up that this flame-haired beauty had sex on her mind at all. We ended up not doing it until our second date, after which she whispered in a cooing voice that she indeed liked my cock. Modern women say this to men they want to keep seeing, I'd come to learn. There would be women who told me *I* was big. This clued me in to what was really going on. Women say these things, I'm guessing, because it must be painfully obvious that we men need to hear it. I recalled green-eyed Gwen in Cordova, and how she was always ready with sweet stroking words. There was a tone of mild surprise to Jill's reassurance, as if she had expected less and was pleased to find more — one of the advantages of living under modest expectations. The whole "its color was its size" thing actually took some pressure off: there was no call to impress there. Yes, I was still on the free throw line, but as long as the ball reached the vicinity of the net, which I could usually manage, I was acceptable.

Soon after our second date, Jill revealed that her previous boyfriend had been black. At the time I thought this was supposed to make me feel good: if he was black and presumably big, and she was happy with me, it must mean I was more than adequate. Unless, of course, her boyfriend was an averaged-sized man, like my friend Raymond, which meant that her comment may not have been a reassurance at all. I never could figure out whether she really liked me or just wanted to save me. Jill and I lasted only a few weeks. I was the one who retreated, an act of confusion as much as anything.

Recalling all this now, I think it's possible I misread the whole situation, including all her words. I squandered the opportunity to be close to an attractive, intelligent woman. I blew it. Sometimes I think it's possible to misread your whole life and virtually everything that transpires in it. To choose one lens and see everything according to its particular distortions. Then one day, if you're lucky, the lens falls off even for just an instant and you realize there are other lenses out there, different shades, opposing visions, even, and that the tale you've been imagining as the story of your life is just that: an imagining, a construction. Beware of the single rendition.

Maybe my Asian-ness played no part in Jill's thinking. Maybe I imagined a tone of reassurance because deep down I was seeking it, expecting it, dreading it for what it implied. Fear has a way of inviting the very thing we dread, like a secret magnet buried deep, drawing toward it shards that would otherwise fall away.

9

Getting Tall

The body contains the life story just as much as the brain.

—*Edna O'Brien*

I still have the pull-up bar that I used as a boy, the kind with the rubber tips that attached at the top of a door opening. Whichever doorway I chose, much grunting took place there. My personal best was thirty repetitions, palms out, face red, veins bulging in surprising places. My spine lengthening. That was the main reason I had asked my father to buy the bar for me, so that I could hang on it for ten to twenty minutes a day, and let gravity pull down my body, stretching the soft tissues and enlarging the spaces between vertebrae. I had read somewhere that it was an effective method. The larger those spaces, the taller I would be, and I entertained visions of myself stretching into the Kareem Abdul Jabbar of the family.

I maxed out at five foot seven. By the time I reached college,

I had accepted my fate and begun using the pull-up bar to hang plants. My handsome white roommate at the time, who was five foot nine, tried to reassure me that it wasn't my fault that I was short. "You're Asian," he said. "What did you expect?"

How did the people of Asia become associated in the Western mind with smallness? Not just with meager body parts, but smallness in general? The answers to these questions followed several paths, but all would lead to the notion that whatever realities existed in the past are today undergoing convulsive change. The first may soon be the last. The short may soon be tall.

The majority of Asians in the United States are foreign-born, and most are shorter than the American average. This certainly helps to perpetuate the association of Asians with small size. But the legacy of smallness goes back a long way—as far back, you could say, as the beginning of the European conquest of Asia.

The people of the East, we recall, were over the course of several centuries trounced by the ascendant people of the West. If not outwardly trounced then forcefully seduced. Force and conquest were male endeavors, outgrowths of the masculine will and direct extensions, some contend, of the masculine appendage. Maybe no modern leader has drawn the connection so directly as President Lyndon Johnson, who, according to biographer Robert Dallek, once responded to persistent queries from a reporter on why America was in Vietnam by unzipping his fly, pulling out his penis, and declaring, "This is why."

You have to wonder how many acts of war committed by kings, conquistadors, generals, and heads of state throughout history can ultimately be explained in the same way. *This is why.* The penis as the symbol of power. The larger the penis, the more power to be wielded. The more power wielded, the larger the penis. Domination was the blood of erection. On the other end, surrender and submission have always been associated with femininity, and females in most mammalian species tend to be smaller than males. Thus smaller size and submission were naturally linked in the Western mind. Acquiescence produced a shrinking effect. Defeat made people small by definition.

So Easterners were small in geopolitical stature long before most in the West had ever set eyes on them. And many Americans' first glimpse of Asian people corroborated the rumor of smallness, or at least of shortness. The first waves of Asian immigrants to America in the 1800s were predominantly Cantonese from southern China, and others from Japan and the Philippines, regions whose people were shorter — in some cases significantly shorter — than the average American, who was at the time the tallest on the planet. The most numerous arrivals were Chinese men, who averaged four feet ten inches, many of them with slight builds. Compounding the impression were Cantonese styles of dress: men wore their hair in queues, long braided strands that hung down the middle of the back, and some donned long silky gowns, the preferred garb of Chinese males since long before the time of Kublai Khan. Many of these Asian migrants to the American West, furthermore, ended up doing what was at the time called "women's work" — laundry, child care, cooking, housecleaning, and gardening — because they were often excluded by whites from traditionally male forms of labor, and because Chinese men did not have the great aversion to such work that white men did. The migrants were willing to do whatever it took to survive.

Then there was the demeanor shown by many Chinese, a temperament that could be characterized as placid, molded by millennia of the traditional Eastern ideals of humility, modesty, piety, harmony, and deference to authority. The rapscallion characters of America's Wild West often interpreted this as timorous.

Never mind that Chinese migrants did some of the most physically strenuous and dangerous jobs on the frontier. These included mining, dam building, land clearing, and construction, tie by tie, of the transcontinental railroad. The white managers of the transcontinental rail project found the Chinese to be the most dependable and hardest-working men they'd ever known. Management initially hired only a small number of Chinese but went on to hire many more after seeing what they could do. Of the ten thousand laborers who built the railway between Sacramento, California, and Promontory Point, Utah, nine thousand were Chinese, whom managers characterized as less demanding, more dependable, pound for pound stronger than whites, tireless, and brave to a fault. One of their primary tasks was to set and ignite explosives, often under dangerous conditions — deep in a cave or far below the edge of a cliff. An estimated one thousand Chinese died while earning much less money than their white counterparts.

These daring exploits did not spoil the popular image of Chinese as weak and effeminate, a view that would take a tenacious hold on the Western imagination. These men with their slight builds and long queues and silky gowns were frequently compared to white women in the American press, and in comics and dime novels and plays through the nineteenth and early twentieth centuries.

Americans took to the image to soothe a chronic insecurity of their own: white American masculinity, unmoored from the Old World hierarchies, was under perpetual threat. "At the grandest social level and the most intimate realms of personal life, for individuals and institutions, American men have been haunted by fears that they are not powerful, strong, rich, or successful enough," writes Michael Kimmel in his seminal work *Manhood in America*. "They have been afraid of not measuring up to some vaguely defined notions of what it means to be a man, afraid of failure."

The image of the ninety-eight-pound Asian weakling assured white Americans that a whole other race of men would always be beneath them, available to be bossed, stepped on, overpowered. To be bigger than.

As the image became entrenched in popular culture, more evidence of Asian diminutiveness came to the fore, mainly through military engagements on the other side of the globe. The United States in the twentieth century fought wars in the Philippines, Japan, Korea, and Vietnam. Soldiers returned home with tales of child-sized people who were nevertheless "tough little fuckers," as one GI put it. A veteran of the U.S. war against the southern Philippine people known as the Moros, Captain C. C. Smith testified that "in hand-to-hand combat, our soldiers are no match for the Moro." They were tough in spite of their size and primitive weapons, often swords and clubs. They were tough because of factors not so measurable.

"If we have to fight, we shall fight," said Ho Chi Minh, the North Vietnamese leader who inspired an army of peasants in flip-flops to persevere against the United States military machine. "You will kill ten of our men, and we will kill one of yours, and in the end it will be you who will tire of it." Ho Chi Minh predicted correctly. His little guys outlasted the big French *and* the bigger Americans. Ho Chi Minh, in his flip-flops, stood all of four feet eleven inches.

• • •

Are all Asian people small, and have they always been so?

The answer to both questions is no—a fact commonly known among educated Asians and Westerners who have traveled widely through Asia. It's worth noting that Marco Polo, who gave detailed descriptions of the most minute aspects of the people he encountered—including even the length and shape of their fingernails—never characterized the inhabitants of Cathay as notably small or short. He frequently extolled their cunning and ferocity and overwhelming effectiveness as warriors, which in Polo's eyes may have elevated their stature.

Conversely, in Chinese historical accounts of first encounters with Europeans, there were no widespread observations of the Europeans' great size. Frequently noted were their beak-like noses, their deep-set eyes, their curious and often repulsive hirsuteness, their range of hair colors, their strong body odor and general filthiness. But nothing about how much taller they were than Chinese. This could be because, in the many centuries prior to the Industrial Revolution, Europeans were not that much taller, and in some cases were shorter, than the Asian populations they encountered. The great divergence in height, researchers theorize, began during the period of China's decline and Europe's ascent, with the gap widening in the eighteenth and nineteenth centuries.

The people of southern and southeastern Asia have long tended to be shorter than people from northern, central, and western Asia, who were often comparable in stature to people of the West. This is particularly true in comparison to southern Europeans (such as Marco Polo, a Venetian), who were generally shorter than northern Europeans. Northern and central Chinese were people of substantial height, an observation documented in many historical manuals. The Terracotta Warriors of

Xi'an, said to be close to lifelike in their dimensions, suggest that ancient Chinese soldiers averaged around six feet in height, tall for any group in the third century BCE. The legendary Chinese warrior Guan Yu and the great eunuch admiral Zheng He were both said to be nearly seven feet tall. Even taking into account inconsistent units of measure and probable historical exaggeration, both men are realistically believed to have been well over six feet tall.

Today, the giant men of the Chinese national basketball teams, whose centers are among the tallest in the world, almost all come from northern and central China. The former Houston Rockets standout center Yao Ming is seven foot six, which even among tall nationalities is aberrantly tall. But less aberrational are the heights of his parents: father, six foot nine; mother, six foot three. (Yao Ming's wife — also from northern China — is six foot three as well.) Up until 2009, both the tallest man and the tallest woman in the world hailed from northern and central China: Bao Xishun, seven foot nine, and Yao Defen, seven foot eight. Yao Defen was recognized as the world's tallest living woman up until her death in late 2012. The tallest woman on record, Zeng Jenliang, who died in 1982, was eight foot one, from China's northern Hunan Province.

Anecdotal records indicate that, during the time of the first waves of Chinese migration to America, men of northern China averaged about five foot seven, with a fair number exceeding six feet. This would have been roughly equivalent to the height of white male conscripts in the U.S. Army and many European immigrants of the time. Had northern Chinese been the first Asian immigrants to the United States instead of the southern Chinese, the American perception of Asian men would have formed along radically different lines.

Imagine if Pacific Islanders—usually lumped in with Asians in the schemata of the three great races—had been the first "Mongoloid" group to migrate in large numbers to America. The Polynesians of Tonga, Tahiti, Samoa, and the Cook Islands, who are genetically linked to Southeast Asians, have long been recognized as among the largest people on the planet. And they were fierce warriors, accustomed to bloody hand-to-hand combat. Had these Polynesians landed on the West Coast en masse in the 1800s instead of the Cantonese, perhaps Westerners would have imagined Asiatics as a gargantuan species of man. What would the rapscallion white cowboys of the Wild West have done then?

Asia and the islands off its coast make up nearly a third of the Earth's landmass, with, today, 4 billion inhabitants. East Asia alone comprises too large a chunk of Earth, too heterogeneous a population, to be understood in simplistic terms. And China alone recognizes fifty-six distinct ethnic groups within its boundaries. The largest of these, the Han, can be broken up into at least eight subgroups, a number of which can be further broken up into hundreds of sub-subgroups according to lineage and geography. Each group sees itself as separate and can often point to distinguishing physical markers.

People of the Asian sphere exhibit a range of physical variation matched only in Africa: from the charcoal-skinned aborigines of Southeast Asia known as Negritos, who share physical characteristics with sub-Saharan Africans, to the fair-skinned, thin-nosed Ainu of Japan, who resemble Europeans; from the lanky golden-hued villagers of Jilin and Liaoning and Shandong to the shorter-limbed Yupiks of Siberia; from the coffee-complexioned farmers of Vietnam and Indonesia to the swarthy, bearded, narrow-faced tribesmen along the ancient Silk Road that connected Cathay to Rome. The road that opened East-West trade also paved the way for a feverish centuries-long intermingling of genes over four thousand miles of wildly changing terrain and temperature.

(Land and climate both affect how bodies evolve. For instance, inhabitants of colder climates generally have greater body mass than people of warmer climates. More mass generates more internal heat, which aids survival in low temperatures. The tallest people of Europe, Asia, and North America live along a northern zone where temperatures are cold but not frigid. Frigid temperatures produce a reverse effect. Long bodies and long appendages allow too much body heat to escape, and so the inhabitants of the circumpolar region, such as the Inuits and Yupiks, evolved to be slightly shorter with extremities to match.)

A passage I stumbled upon in a footnote of a translated Soviet-era text gives a physical description of a man living along the old Silk Road in a region called Xinjiang in China's remote northwest: "He is medium tall with pale skin, torso thin and strong, hands large, face with Indo-Iranian qualities mixed with Mongoloid. The eyes are narrow, light brown to gold in color."

My first thought: he is Asian.

So, short Asians? Yes. There have been many; we are still many. The first to arrive in America, as I've said, were short. Most of the subsequent generations immigrating to America were shorter than the American average, but the same is less true of their children and grandchildren, and of the most recent waves of Asian migrants. American-born Asians have inched upward. And average height among Asians in Asia is inching up too, in some cases shooting up.

The average height of a population can change dramatically even over brief periods of time, according to researchers in the field of human physical growth known as auxology. Auxologists say that great height variation already exists within all major racial groups, but the potential for bursts of height increase may be greatest among people who live a long time under conditions of deprivation and then enter a new phase of prosperity. This describes much of Asia.

Auxology is a cross-disciplinary field involving biologists, anthropologists, economists, sociologists, historians, physicians, nutritionists, and other specialists. Auxologists accept that height is determined by the interplay of nature and nurture, but they tend to focus more on the various aspects of nurture. A main precept is that average height reflects a society's overall well-being. More social stability, more equal distribution of wealth, more nutritious diet, better health care, better social services, less illness, less pollution, and less physiological stress all lead over time to greater average height. This is why people in developed countries tend to be on average taller than those in still-developing ones. Studies show that the children of privileged families in all racial groups, who uniformly lead healthier lives, tend to be around the same above-average height.

War, instability, famine, and disease are great height reducers, which explains why many Asian populations today are, according to some estimates, shorter than their ancestors in Asia's golden age. Asia has suffered precisely these height-reducing upheavals over the last millennium, many in the past several hundred years, and some within the past fifty years. Deeply entrenched poverty, deprivation, and instability still plague a number of Asia's developing countries.

Famine alone repeatedly devastated Asian populations, with China bearing the worst brunt. A single famine in the 1300s killed 6 million Chinese. A series of famines in the 1800s killed 45 million Chinese. Seven of the ten worst famines of the twentieth century took place in Asia (or ten of ten if you count the former Soviet Union as primarily in Asia), four of them in China. The worst recorded single famine in human history killed more than 40 million Chinese in the late 1950s and early 1960s, during Mao's catastrophic attempt to turn China overnight from an agrarian to an industrial state.

The numbers generally do not include those who died of drought, disease, and war, which often coincided with famine. And they do not reflect the hundreds of millions who suffered the effects of famine but did not succumb. One effect was stunted physical growth. Amid the living hell that China must have been for survivors, reaching maximum height was likely not high on the list of priorities.

The southern Chinese who began migrating to America in the nineteenth century did so partly because of a long seafaring tradition in the southern coastal region. And these migrants tended to be short, slight people to some degree because they and their recent ancestors had been ravaged by centuries of tumult. They were human rags who had survived an apocalypse, and who were as much escapees as migrants. They found themselves, as my family did in the 1960s, in a land of giants.

In 1850, Americans were the tallest people in the world, and they remained so for most of the twentieth century, reflective of their socioeconomic rank. Today, American men rank ninth and American women fifteenth in average height. It appears that other nations, according to auxologists, have learned to take better care of their citizens. The tallest people today come from northern Europe, and the tallest of the northern Europeans are the Dutch. The average height for Dutch men today is six foot one, three and a half inches taller than the average for American men.

I traveled to Amsterdam not long ago and found myself craning upward at bellmen, restaurant hostesses, store clerks, and ticket punchers. The concierge at my hotel, a jocular fellow named Dedrick, whom I would come to call The Ded, was the same height as Larry Bird, six foot nine. A few times I returned to my hotel to find The Ded chatting it up with the doormen, who were almost as tall as he was. Walking into their presence, I felt like Bilbo Baggins joining an NBA huddle. My high five was for them a low five.

Researchers cite the Netherlands' relatively equal distribution of wealth, its high incomes, its exemplary prenatal and postnatal care, the overall quality of its health and social services, and, of course, the protein- and nutrient-rich diet available to almost everybody. Milk and meat for all. The story is more remarkable for the fact that a mere century and a half ago, the Dutch were among the world's shortest people. The average Dutch male in 1850 was five foot four, three inches shorter than the average northern Chinese of the time. At one point, a quarter of the men who tried to enlist in the Dutch army were rejected for not meeting the minimum height requirement.

What happened to the Dutch in the subsequent five generations — when all the good stuff mentioned above began — is occurring now among people in many parts of Asia, particularly where peace and prosperity and a Western-influenced diet have been in place for a few generations.

Some have suggested the Japanese might become Asia's counterpart to the Dutch. In 1950, Japan's average height was the shortest for any industrialized country. But after only sixty years of political stability, relative wealth, improved health care, and a Western-influenced diet rich in protein, the average height has shot up nearly five inches to five foot seven for men. Walk into a high school in any of Japan's large cities today and you'll find the same big-footed, lumbering, and often towering teenagers you would meet in American schools. The Japanese "could equal American height standards in the next generation," says Richard Steckel, an economist and anthropologist at Ohio State University and a leading auxologist.

A similar extended spurt is occurring among South Koreans, whose average height compared to North Koreans' provides what some consider irrefutable proof of the impact of environment on physical stature. Koreans were a homogeneous people prior to the partitioning of the country at the end of World War II. Northern Koreans may even have been taller by a hair.

The height divergence began when the two economies diverged, with the greatest disparity coming among Koreans whose peak growth years took place in the mid-1990s. At that time, as South Korea's economy soared and its people thrived, North Korea experienced a famine estimated to have wiped out 10 percent of its population, or 2 million people. The UN found that chronic malnutrition left four out of ten, in some cases six out of ten, North Korean children physically stunted. Anthropologists who measured North Korean refugees were stunned to find that most teenage boys stood under five feet tall and weighed less than one hundred pounds. North Korea's soldiers tend to be shorter than their South Korean counterparts, but North Korea posts its tallest men along borders to give the impression of size. Outsiders who visit North Korea "are often confused about the age of the children," wrote my then-colleague Barbara Demick of the *Los Angeles Times* in 2004. "Nine-year-olds are mistaken for kindergartners and soldiers for Boy Scouts."

In prosperous South Korea, the average seventeen-year-old boy stands five foot eight, just slightly shorter than the average American boy of the same age. South Korean teenagers over six feet are becoming increasingly common. I've observed the same in cities all over Asia, even in still-developing countries like the Philippines and Indonesia. The marked size differences between generations can be visually striking, with children growing a foot to a foot and a half taller than their parents.

I recently sat inside a Pizza Hut in Quezon City during an all-you-can-eat night and watched an immense Filipino boy ingest two pepperoni pizzas, minus once slice each for his diminutive parents. The boy ate the pizzas like big cookies. The mother, who looked as if she had just left the rice fields, was the size of one of his legs.

I've seen the same contrast among Asian families in the United States, with American-born and raised children towering like giraffes over immigrant relatives. In my own large extended family, it's a running joke how much taller the younger generations are compared to the older. Standing side by side with his grandmother, my nephew Josh could use the top of her head as an elbow rest. He could rub his chin on the top of my head if I let him. In group photographs, the younger generations seem almost to constitute another species, an observation my father and I once made about Americans when we first arrived stateside. In this land of giants, our children and grandchildren are becoming gigantic.

As fortunes change, bodies tend to change too. Almost nothing about average physical dimensions is fixed. We mammals, it turns out, are quite elastic. And as bodies change, fortunes will likely improve. Taller, sturdier, healthier Asians will undoubtedly benefit from catching up with other racial and ethnic groups. (They will likely also face the downside of bigger bodies, such as clinical obesity, weight-related joint issues, and high blood pressure and its attendant problems.)

Cultures everywhere and throughout history pay homage to the bigger man. Vertical size, however unfair (and inaccurate), translates in many minds to capacities in other areas. Height communicates power and potential, and thus bestows status. Short stature is the simplest and most convincing explanation for the difficulty experienced by many Asian men in the Western dating and mating scene. In a lucid 2010 study published in the journal *Economics & Human Biology*, researchers documented what most of us instinctively knew, that women almost everywhere prefer taller men. A five-foot-four-inch man of any ethnicity will face romantic challenges in a country where half the women are taller and more than a few tower over him. It's that simple, and that brutal. Height may also help explain why so many talented and driven Asians hit a bamboo ceiling in the corporate world: because those deciding who gets promoted tend to be taller than average and tend to prefer people like themselves.

As fortunes and bodies change, so will our mental associations. I foresee a time, maybe just a generation or two away, when the easy coupling of the word "little" with the word

"Asian"— *There was this little Asian guy. . . . I met this little Asian chick. . . . Hello, my little Asian friend* (a greeting I heard often from one colleague) — may not be uttered so easily. Word pairings will change, associations will shift and reconfigure. The fact of the short Asian, and the popular image of the short Asian, could someday become as anachronistic as the belief that state communism works or that women and dark-skinned people cannot lead.

10

Wen Wu

Every man is a quotation from all his ancestors.

—*Ralph Waldo Emerson*

No one can be free who has a thousand ancestors.

—*L. M. Montgomery*

The three dead men were my ticket to China.

They were discovered in the back of a newly offloaded cargo container at Terminal 18 on Seattle's Harbor Island in the winter of 2000. Cause of death: dehydration, starvation, and exposure. Young men all. Others in the container survived. The group had been in transit for sixteen days across the Pacific Ocean. They had no food or water, except what they had carried on board and quickly consumed, no toilet, no windows or light, and barely any airflow. In pitch darkness, they sat or lay on the floor, and were tossed around when the waves got rough. Many became seasick. The air was thick with the stench of vomit and sewage. Several others were approaching death when their container reached Seattle.

Chinese "stowaways" had been arriving by the hundreds on the West Coast since the last months of 1999. It became a national story. Americans were astonished by the desperation of the stowaways to sneak into the country. When authorities revealed that the three who died had come from the Chinese province of Fujian, I immediately hatched a plan. I was working as a journalist in Seattle, covering issues related to minorities and immigrants. I proposed to my editors that I travel to Fujian to find out why people there were willing to risk so much to come to the United States. It was a legitimate proposal.

In the end, the answer I found was a version of an old, familiar one, the reason behind almost all human migration: to escape a fixed destiny (usually involving poverty) and to forge a new one (usually involving a dream of wealth).

But there was another reason I wanted to go to Fujian, unrelated to the stowaways. A personal reason, tied to my own secret self-education. Fujian was the place from which a fifteenth-century Chinese admiral named Zheng He had launched his epic voyages. Most Americans had never heard of him, and information about him in English was hard to come by at the time. I knew only a little about him myself, and wanted to know more. I had a feeling his story would bring my investigation of Asian manhood to a new level of revelation, and I had been seeking a work-related reason to travel to that part of the world. My first choice would have been Yunnan province, where Zheng He (pronounced *jung huh*) was born, but Fujian, the home port for his fleet, was almost as good.

My curiosity about the admiral was part of a search for a

beginning to my own story, one that went farther back than my parents and a ticket to America. I had come to feel like an amputee of sorts, severed from my ancestors — in the Philippines and the Malay Peninsula, and before them, my ancestors from China. If you trace the history of any kingdom or country in Asia, particularly East Asia, you will eventually come to the story of China. The Middle Kingdom was the source from which East Asia flowed and formed, its traditions shaping much of the various cultures of the continent. I knew there were aspects of myself shaped by this outflow, and that I had forefathers who could teach me something. The admiral, I sensed, was one.

"You know Zheng He!?" said Li, my guide and translator in Fujian. He was a gray-haired man with cataract eyes and a mostly stoic disposition that lifted when a topic interested him. Then his eyes lit up and his eyebrows danced. He became chatty, and he would end his sentences with a little laugh, almost a giggle.

By way of answering, I pulled out a creased magazine article about the admiral that I had been carrying around in my back pocket. I offered it to him, but he only glanced at it and waved his hand.

"Oh, I see," Li said. "Zheng He was big man. Like Columbus. You know Columbus? In fourteen hundred ninety-two. . ."

". . . Columbus sailed the ocean blue," I said.

"Yes, yes. Zheng He was Chinese Columbus."

"I know," I said.

"Except one difference," he said with a glint in his cataract eyes. "Zheng He sailed long time before Columbus. Zheng He was first." Li giggled.

Fujian sits on China's southeastern coast, a rugged, sloppily beautiful place on the edge of the Taiwan Strait. I visited the port, near the capital city of Fuzhou, where the admiral launched his magnificent fleet, and I tried to imagine what that scene must have looked like in those days. Fujian had been China's version of the Wild West going back centuries. A mountain range that acted as a natural barrier allowed the province to develop independently, evolving into a land of ruffians and schemers and cutthroat entrepreneurs out to make a quick fortune. There were pirates in the mix, too. Fujianese were known as physically strong and combative, but also industrious and tenacious beyond rival. Chinese cite these characteristics to explain why, at the beginning of the twenty-first century, more than half of Asia's four dozen billionaires of Chinese ancestry came from Fujian or descended from Fujianese.

Li brought me to shrines dedicated to the admiral and to the homes of several local historians. I took notes; Li translated and frequently added his own two cents. He was right about Zheng He preceding Columbus, but the span between the two explorers was not that wide. From a long view of history, you could almost call them contemporaries. The Chinese admiral completed his last voyage about sixty years before Columbus anchored in the Caribbean.

Because so much of Zheng He's story was destroyed or left untold for so long, much is still being pieced together, with fragments often coming from the foreign lands to which he traveled. New information about him surfaces every few years. What's been known for some time is that he was born into a rebel family in what was then a Mongol province in Central Asia (now part of China), was captured and ritually castrated by Ming soldiers when he was a boy, and forced to be a servant in the court of a powerful prince.

As is so often the case, people who go on to become great

do so to some degree because of sheer luck. Zheng He's first stroke of good fortune was that his patron prince was a benevolent man who grew to like him, and allowed him to become educated in literature, philosophy, and the art of warfare. Zheng He developed into a brilliant and physically imposing military commander. His second stroke of luck was that his prince became emperor. One of the new emperor's pet projects was the launching of the greatest navy ever assembled. The emperor gave the helm of this navy to his servant-turned-admiral, Zheng He.

The Treasure Fleet was larger than all the navies of Europe combined. It consisted of 28,000 soldiers and merchants on three hundred vessels, including supply ships, troop transports, combat junks, patrol boats, and a couple of dozen tankers to carry potable water. The largest of the vessels, the *baochuan*, or "treasure ships," were 400 feet long and 170 feet wide, and had nine masts, red silk sails, multiple decks, and luxury cabins with balconies. They were by far the largest wooden ships ever built, and were the most technologically advanced vessels on the planet.

From 1405 to 1433, Zheng He led seven expeditions that reached as far as present-day Iran, Kenya, and Tanzania. Military conquest wasn't an objective, but the admiral let loose his cannons when necessary. In what is now Indonesia, he took great pains to avoid a confrontation with pirates, but when the pirates attacked, Zheng He responded by killing five thousand of the marauders and capturing three of their ringleaders, who were later beheaded. The fleet used military force in Arabia, East Africa, and present-day Sri Lanka, usually with the Chinese drawn in reluctantly. All enemies eventually bowed. But confrontations were few. Most would-be challengers would acquiesce at the mere sight of the fleet.

By the end of the seventh voyage, Zheng He and the Treasure Fleet had explored parts of two continents (three if indeed the fleet reached Australia, as some claim), introduced Chinese goods to a vast swath of lands, and collected tribute from more than thirty-five kingdoms. The admiral covered more distance and spread his homeland's culture farther than any European seafarer at the time.

Why hadn't I learned about him in school? Why does the book on my shelf, the one I've owned and read since my high school days, *The Explorers* by Richard Humble, about the world's great maritime adventurers, not mention Zheng He even once?

The answer, to some degree, is that the Western world in which I was raised did not view the story of Zheng He as germane. It was the same reason why I had not been taught in any meaningful way that the great empires of Asia were, not that long ago, the preeminent civilizations in the world. Is it far-fetched to imagine that knowing this — and knowing that my white teachers and classmates also knew it — would have made a difference in the way I viewed myself as a boy, as a young man? That it would have made a difference to many like me? Some of my friends, including numerous Asian brothers and sisters, think it is indeed far-fetched — and inconsequential. "Who cares?" they say. "It was six centuries ago!"

Knowledge of history, I keep learning, affects people in vastly different ways. In my own life, history always had a profound presence, planting the seeds of ideas that constantly changed the outlines of my worldview. I grew up with the unquestioned belief, based on the history I learned in school, that the great explorers and conquerors of humankind were all European. The visage of noble power was European. The face of the trailblazer was a European face. *Not* to be born with that face was to be destined to a lesser fate.

Learning that Zheng He existed, that he did the things he did, expanded in me the idea of my own potential. Part of what I sought in the quest for a beginning to my story was a sense of a path both behind and ahead of me, one that offered a vision of what was possible for a man of my lineage. Just as Lapu Lapu whispered to me on Mactan Island that it was in the blood of our people to be capable of defeating a stronger enemy, I learned from Zheng He that Asia could produce fearless, questing men who were not tyrannical or bloodthirsty, and who actually contributed good things to the world

Zheng He's voyages were not a fluke but a culmination of centuries of innovation and progress. In the millennium prior to the Treasure Fleet, Asia — with China and India as hubs — was by far the most advanced, cultured, and commercially vibrant continent on the planet. Asia contained the world's five largest cities, all at the heart of great empires and all interconnected by a bustling network of trade routes. From the time of Zheng He up until the early 1800s, China and India together accounted for half of the global economy.

Asia was the epicenter of innovation. Some scholars, including the renowned Cambridge scientist and historian Joseph Needham, have speculated that without three Chinese inventions — paper and printing, gunpowder, and the magnetic compass — Europe might not have come to dominate the world when and how it did. What a novel concept, that Europeans rose to preeminence in part by stepping on the shoulders of Chinese innovators. Of course, it's impossible to pinpoint the origin of an idea or to know the capillary routes through which knowledge traveled. But just the notion of Eastern influence on Western ascendance would have been an affirmation to me as a boy seeking historical fathers and looking for signs of ancestral involvement in the game. Perhaps I would not have felt so far removed from the story of humankind.

When Li wasn't driving me around or showing me the sights around Fuzhou, he was reading in his car, glasses dangling precariously on his nose. He went through two or three books during our time together. He would get so absorbed that my appearance at the car window, after I'd gone off on a short sightseeing jaunt, would startle him. It would take him a moment to remember who I was. "Yes, yes," he'd say. It turned out that Li had multiple college degrees and was well versed in, among other things, horticulture, local history, and Confucian philosophy.

"Zheng He was powerful man, not violent man," Li told me over tea on the islet of Langqi, at the mouth of the Min River. This was his response to my recounting the story of how the admiral, on one of his expeditions to India, had tried to avoid a fight with the notorious pirates in the Strait of Malacca. The admiral had collected intelligence on the pirates and decided to take a different, longer route to avoid what would certainly be a violent confrontation.

"Do you think Zheng He was afraid of the pirates?" I asked, wanting more to get a reaction than an actual answer. I figured he couldn't know the real answer.

"No, no," Li said, giggling. "Not quite right." He took a sip of his tea and appeared to be collecting his thoughts for a definitive response. His eyebrows rose and lowered. His gaze grew distant for a few moments. Then, as if deciding that the matter was too complicated, he simply sighed and repeated, "Zheng He powerful man, not violent man." At some point he used the term *wen* to describe the admiral, but I wasn't familiar with the term at the time, and Li could not explain it in English. Eventually, after I had left Fujian and returned stateside, I think I figured out what Li was getting at.

It was the concept of *wen wu*.

Coming to understand this idea unlocked the door for me to a new insight on what it means to be a man in Asia. *Wen wu* represented the ideal man, the Perfect Man, or in any case it expressed a more complex, and complete, idea of manhood. According to the Confucians, anyway. This ideal man was hard to find and harder to become. It took a lifetime's work, and almost all fell short, but the vision was clear on what to strive for. In East and Southeast Asia today, this notion of perfect manhood is still what many mothers and fathers wish for their sons, what countless boys and young men wish for themselves, and what traditional women long for in a lover or husband.

For the past two thousand years in China, you could not be merely a tough guy to be considered an ideal man. You also had to be scholarly, poetic, and wise. The manliest of men were philosopher-warriors, and more philosopher than warrior. A cultivated mind was more highly esteemed than big biceps or deft swordsmanship. The man who could recite Confucius's *Analects* from memory was considered more substantial — and more desirable — than the one who could lick any man in the room. The ideal man could do both.

Thus the way to perfection took two paths — *wen* was one path, *wu* the other — and an aspiring ideal man traversed both simultaneously. There's no precise English equivalent. *Wen* is defined as "literary and cultural development" attained through study, creativity, and reflection, according to Kam Louie, a leading scholar of Chinese masculinity studies. *Wu* is often summarized as "physical strength, fighting skill, and the capacity to wage war." To put it simply: *wen* refers to the development of the mind, *wu* to the development of the body.

The order of the words matters. A classic of the Confucian tradition, *Spring and Autumn Annals*, puts it succinctly: "The virtues of *wen* are superior, the greatness of *wu* is lower, and this has always been and will always be the case." The softer, cerebral man is of a higher order than the harder, martial man. The scholar trumps the soldier. The perfect man is the brilliant scholar who also happens to know the fine art of throwing a roundhouse kick to an opponent's head — but figures out a way *not* to have to throw it.

The noblest application of *wu*, when tempered by *wen*, was the containment rather than the use of power. Uncontrolled or undisciplined violence was seen as an expression of weakness. Restraint was the ultimate example of strength. Asian martial arts, as taught by the ancients, trained the ability to hold back, when possible, to disable rather than kill an enemy. This holding back also applied to inner appetites. For example, careless emission of sexual energy was discouraged. The containment of excessive sexual urges was held up as a virtue.

As with so many other aspects of life, the Chinese heavily influenced their neighbors. The Japanese adapted *wen wu* to form the concepts of *bun* and *bu*. The samurai sought to be educated and cultured as well as proficient at slicing and dicing enemies with their swords. Medieval Japanese narratives refer to the ideal of the educated poet-swordsman. Other elite fighting units in East Asia, such as the Sulsa warriors of Korea and the Shaolin monks of China, all considered spiritual and intellectual development more desirable, and more important, than pure martial skill.

The Chinese general Sun Tzu was an accomplished military strategist but was a thinking man first, a fighting man second. He is known best for his classic philosophical treatise *The Art of War*, which today is required reading in military academies and business schools around the world. A sampling of the general's *wen wu* philosophy: "To win one hundred victories in one hundred battles is not the height of skill. To subdue the enemy without fighting is the height of skill."

This isn't to say that *wu* men, those get-it-done alphas who overpower rivals, did not exist in Asia. They were legion, some known in the West more as murderous invaders than skilled conquerors.

Attila the Hun, whom fifth-century Romans would come to call the "Scourge of God," descended from the nomadic peoples of the Mongolian steppe. He terrorized lands as far west as present-day Italy, Greece, and Germany. The Greek historian Priscus described Attila as "short of stature, with a broad chest and a large head; his eyes were small, his beard thin and sprinkled with grey; and he had a flat nose and tanned skin, showing evidence of his origins." His army of a half-million horsemen with tattooed faces and great manes of knotted hair created a fearsome sight. Entire regions would empty of inhabitants at the rumor of his approach.

In the thirteenth century, Genghis Khan, another Mongol, carved out the largest contiguous empire in history, from what is now Korea to the eastern edge of present-day Poland. His army conquered more lands and people in a quarter century than the Romans did in four centuries, redrawing the boundaries of half the known world. (Famous for his rapacious sexual appetite,

Genghis may hold the distinction as the most prolific man in history. Geneticists have found that about 16 million men from China to the Middle East—or about one of every two hundred men on Earth—are his direct descendants.)

A macho tradition has always existed in Asia, but in China the ethos has seldom remained dominant for long periods, according to Kam Louie. What distinguishes the Chinese from other societies with dualistic ideals of manhood is their level of loyalty, their tenacious adherence over millennia, to *wen wu*.

Zheng He was a *wen wu* man. This is what I think Li was trying to tell me over tea on Langqi. Zheng He had a cultivated mind, a strong body, and a hand on the helm of the world's largest navy. He had the power to conquer wherever his Treasure Fleet went. Instead, the admiral and his men showed remarkable restraint, conducting themselves in a wholly different way than did the great European explorers who entered the world stage shortly after. Rather than subdue with force, Zheng He and his fleet won over the locals with goodwill, generosity, and just enough of a show of strength to garner the appropriate respect. This was the use of "soft power" centuries before Harvard political scientist Joseph Nye coined the term. In many lands visited by the Treasure Fleet, Zheng He became revered for his benevolence and even came to be deified.

One gets a sense of the admiral's personal philosophy from the few words of his that have survived. A month before the Treasure Fleet's maiden voyage, at the age of thirty-four, Zheng He commissioned an epitaph inscribed on a stone pillar over his father's grave in Yunnan province. He worshipped his father, who had died in battle. The epitaph, one of only three known testimonials from the admiral, described his father's character:

He was content as an ordinary commoner, but he was brave and decisive in his ordinary life. There was no one in this community who did not look up to him. When he encountered the unfortunate, including widows, orphans, and others with no one to rely on, he routinely offered protection and aid. He cherished the bestowal of extraordinary favors. By nature, he was fond of doing good.

This revelation of a softer version of manhood as the ideal in much of Asia provided another piece of the answer to the question of how Westerners came to perceive Asians as less masculine.

The Chinese men who migrated to America in the nineteenth century, in addition to being shorter, were steeped in the ideals of *wen wu*. These men journeyed to "Gold Mountain," as America was called, largely unaware that they were entering a world with radically different notions of manhood, notions that were being re-forged at the time by the ultra-macho dudes of the Wild West. Mark Twain called them real men, "not simpering, dainty, kid-gloved weaklings, but stalwart, muscular, dauntless young braves, brimful of push and energy, and royally endowed with every attribute that goes to make up a peerless and magnificent manhood."

The badass cowboy-soldiers of the West made an assessment of the Chinese with their gowns and braided queues, their reserved demeanor and soft-spoken ways, and they came to their conclusions: Polite and courteous? Studious? Trained in the quieter disciplines? Refined in their mannerisms? Willing to do humble work? Not interested in random, fleeting sexual encounters? Unwilling to brawl and kill over minor disagreements? Taking a detour to avoid a fight? *Wimps, that's what they are.*

This interpretation of Asian *wen wu* behavior persists today. As of 2013, roughly two thirds of Asians living in the United States were born abroad, and the males very likely grew up with some version of *wen wu* as the abiding principle of manhood.

I see this principle playing out in the strong penchant for study and educational achievement, so denigrated, so parodied among non-Asians. Walk into the library of any major university, and you'll likely find Asians — male and female — taking up a good portion of the study spaces. In some West Coast schools, Asian students are known to take over sections of libraries, if not entire libraries. The ideal of *wen*, combined with the traditional Confucian ethics of sacrificing for long-term goals and subordinating personal drives for a greater good, contributes to Asian Americans' attaining on average higher educational levels than any other group, with 50 percent earning at least a bachelor's degree (compared to 30 percent for whites). Asians also have the highest percentage of graduates with a master's, professional, or doctoral degree.

I see it in formal and informal gatherings of Asian men, among whom there seems less of an inclination for bluster and bravado and macho posturing. In countless staff and board meetings and public hearings I've attended over several decades, Asian men and women tend not to be the boisterous, confrontational ones in the room. Even in heated conflicts involving Asians, there's often an element of *wen wu* restraint.

Crime studies uniformly show that Asian Americans are the least likely among ethnic groups to commit violence. There is the occasional loose cannon who makes news, such as Seung-Hui Cho, who killed thirty-two students at Virginia Polytechnic

Institute and State University in 2007. But the fact that he was Korean added to the grotesquerie of the incident. Every crime report generated by the federal government in the past five decades shows that Asians, in relation to their share of the population, are distinctly underrepresented in the ranks of murderers, violent attackers, rapists, and armed robbers. The reputation for peaceful inclinations is borne out in the numbers.

These are sweeping generalities with countless exceptions, I realize. Perhaps I would not be as convinced if the pattern didn't hold true in my own family. I saw glimmers of *wen wu* in my father, though I did not realize it at the time. In his case, *wen* and *wu* were often in conflict, jockeying for supremacy. I grew up with the same inner tension. While my adolescent consciousness, fed by endless hours of American television and movies, pushed me to become Rambo, the muscular man of action, the unstoppable force of nature, something deep inside, cultivated by years of parental browbeating, compelled me to lay low, talk softly, and maintain a killer grade point average.

In Fujian, I had not expected to find a link between Zheng He and the three dead stowaways in Seattle, but the longer I stayed in China and the more I learned about the admiral, the more I began to understand the filamental connection. The full story of Zheng He, in fact, has significant bearing on the subsequent decline of China, which eventually spawned the circumstances that impelled the stowaways.

Here's an encapsulation of six centuries of history, beginning with Zheng He's last voyage and ending with the dead stowaways on Harbor Island:

The Treasure Fleet could have continued south along the East African coast, rounded the Cape of Good Hope, sailed north to

Europe, and gone possibly as far as the Americas. (A handful of maverick historians claim that the Chinese in fact reached the Americas before the Europeans.) But shortly after the 1433 expedition, the emperor who had championed the Treasure Fleet died, and an influential group of Confucian scholars came to control the palace. The Confucians viewed exploration of foreign lands, looking outward, as wrongheaded; looking inward was a greater virtue. They also saw the expeditions as too expensive and the quest for profit as vulgar.

The Confucians dismantled the Treasure Fleet, destroyed all oceangoing ships, and banned further naval exploration, punishable by death. To ensure that no one followed in the admiral's wake, they burned Zheng He's naval records. The admiral and his magnificent fleet were purged from the archives.

Some scholars regard this as the beginning of the end of Chinese preeminence. The Chinese could have led the world but instead chose to leave it, retreating into the folds of their own kingdom. India did much the same. The rulers of these ancient Asian civilizations believed that they had everything they wanted and did not need contact with outsiders. The rest of the world offered nothing they desired. Complacency and a stubborn clinging to tradition prevailed during the same centuries when Europe became enlightened, embraced scientific inquiry, and created technologically advanced world-beating armies. The East stagnated as the West transformed, a centuries-long process that culminated in Europe's domination of Asia and much of the planet.

Cultural lethargy, overpopulation, and too many decades of failed policies led to dire conditions in places like Fujian province, where, as recently as 2000, the average worker earned between $120 and $250 a year. Tens of thousands of impoverished

Fujianese stole onto ships for a chance to start over somewhere else, somewhere better. And so sometime in the first days of the new millennium, Zhang Hui, Jiang Dianbiao, and Zhu Benqing joined fifteen other working-age men inside a dank cargo container on board the freighter *NYK Cape May,* bound for Seattle. The rest you know. It was Zhu Benqing who died last, with the outline of the Seattle skyline in sight of the ship's crew.

"You get what you looking for?" Li asked me at the end of my trip.

"I don't know," I said. And I thought but didn't say, "I think I'm just beginning."

The journalistic story I wrote covered the bases, painted a portrait of a rising superpower with vast numbers of people still waiting for the economic miracle to save them. But the significance of the trip for me was that I had tapped into an essential insight on Asian maleness, and I continue to plumb its depths.

I'm still chagrined that I grew up learning nothing about Zheng He and the greatness of the East that he represented. Eurocentrism was partly to blame, but perhaps more culpable were the Confucians who renounced the admiral and closed the doors of the Middle Kingdom. A small circle of Western scholars knew of the Treasure Fleet as early as the 1930s, but Zheng He's story was not presented to the American public in any meaningful way until Louise Levathes's 1994 book *When China Ruled the Seas.* The story of the Treasure Fleet did not "go viral" among educated readers until a 1999 story by Nicholas Kristof, "1492: The Prequel," ran in the *New York Times Magazine,* the article that I kept in my back pocket during my stay in Fujian.

In other words, prior to the last year of the last millennium—less than fifteen years ago as I write this—Zheng He's voyages were essentially an untold and unheard story among the vast majority of Americans. Most of my friends and colleagues have still never heard of him. In my travels around the country, I would broach his name among college students, Asian Americans included, and they would show the same quizzical expression that my own children gave me at home.

"Zheng *who?*"

The history we learn is always only a partial story. A sliver of the whole. Sometimes you have to go looking for the rest of it. I wonder how many other great questing men — and women — of Asia I have yet to learn about.

11

Yellow Tornado

The thing that falls away
Is myself.

— *Kintsune*

An old friend had sent me the video. Alone one night in the office, I watched it again and again. I had seen the actual event live on television in 2004, but the video let me study, and savor, how it unfolded. Each time, I allowed myself the fear that I felt when it first happened. I feared the yellow man would lose. He'd lose on the grandest stage, the Olympic Games, with a billion eyes watching and the secret hopes of yellow men everywhere dashed.

The 110-meter hurdles required explosive speed and power, the fast-twitch stuff, those equine glutes. It required technique, too, but before and after each hurdle, a full-on stallion gallop. The lineup was familiar, black men from Europe and the Americas. But in the fourth lane, easing into a starting block, was Liu Xiang of China. How did a yellow man make it to the finals of a sprint event? There must have been a mix-up, an accident, an injury. Liu must have been a last-minute replacement, a decorative flourish to contrast with the solid block of black muscle on the starting line.

The gun fires.

Liu bursts out among the leaders. One hurdle, two hurdles, three. He hasn't fallen. He hasn't fallen behind, either. What's this? By the fifth hurdle he's neck and neck at the head of the pack. Six, seven. By the eighth, the yellow man surges ahead, and he's moving so fast my eyes can scarcely keep up, and my mind, too. Nine, ten, home stretch. He pulls away, and his shock of jet-black hair crosses the finish line, an ebony mop dancing in the wind. The time: 12.91 seconds, tying the world record.

The crowd at Olympic Stadium in Athens appears stunned, as if a comet had come from nowhere and streaked across the sky. Some spectators stand with hands on hips, looking around, their expressions seeming to ask, *Did you see that?*

Afterward, Liu talks to the news media. "It is a proud moment not only for China but for Asia and all people who share the same yellow skin color," he says. He is tall, sleekly muscled, with a block jaw and a winsome smile that flashes easily. More "miracles" would come, he predicts. His people could "unleash a yellow tornado on the world."

He went on to break the world record and become a cultural icon in China, a symbol of possibilities. Young Asian men everywhere now had irrefutable proof of the potential in their bodies, proof that it was possible for them to compete, and win, in a sport thought to be the exclusive domain of black and white athletes. I certainly thought that. Most everyone I knew did too but did not say it out loud. The unspoken assumption my entire life was that, in power sports, yellow men need not apply.

It was only a footrace, I know. I'm almost embarrassed by the significance I ascribed to it. Yet somehow it did seem to unlock a tiny shackle inside me. My frame of mind was such that every little thing could be a metaphor for a larger hope. And I must have had a particularly trying day at the office. I was still learning to do my job as a national correspondent for the *Los Angeles Times* (based in Seattle), feeling overwhelmed by the workload and harboring a growing suspicion that I was in over my head. That night I was listless, not wanting to go home but not wanting to leave the office either. My friend's video sat at the top of a pile of mail. I popped it in the VCR, watched it, and found myself buoyed. Without thinking too hard about the reasons, and with no one around to inquire, I hit the rewind button and watched those thirteen seconds over and over, once in super-slow motion, freeze-framing the finish.

Signs of the tornado have sprung up everywhere, and I've been caught up in the swirling eddies. The American public seemed to acknowledge in the 1980s what economists had been tracking much earlier, that a tectonic shift in the world order was under way, a phenomenon Gore Vidal characterized as the awakening of "the long-feared Asiatic Colossus."

I took note in the 1980s, too. My own life was undergoing radical changes at the time. I was older, a little less awkward, and a bit more assertive socially, and I was beginning to tune in to who I was and what I was supposed to do in the world. With this, I began to sense a shifting in me, something that in retrospect I can describe as a rising up from the hole of my ancient inherited shame.

How my climb was affected by the stirrings of the yellow tornado on the other side of the globe I can only speculate. The one true thing I can say is they *felt* connected: my clambering upward coincided with the rising up of the continent that bore me, our climbs seemingly tied by invisible filaments stretched across oceans and continents, by strands of common history and torment, and now promise.

Part of it had to do with the search for identity becoming a cross-border, globalized experience. Immigrants of the past were often permanently cut off from the old country. Now, with the ease of travel and communication, immigrants can remain connected and forge identities that span oceans. So I was both Asian *and* Asian American. And Asia's rise helped me to rise, if only as a source of pride that I could point to and mysteriously tap for extra reserves when my energy flagged. Sometimes it propelled me farther than I thought I could go. Maybe that's why I hit the rewind button: I cross the finish line every time Liu Xiang does.

I'm still not out of the hole completely. Some days I'm dropped back down to the bottom. There may always be a gravitational pull toward the nether depths for me. Nothing inherited is escaped easily, and it didn't help that my first three decades in America were marked by perpetual flux and disorienting fits and starts, my own imperfect storm. These convulsions seemed to intensify even as I made progress. I became my own tornado, spinning lopsidedly, clearing new paths but also wrecking things.

• • •

In between jobs and my travels to Asia and my secret self-education, and trying to live up to dueling grand concepts of manhood with a capital *M,* I routinely failed at being a man on more pedestrian levels.

I got married shortly after graduate school. She was a bright, lithe, and lovely woman, of German and Irish ancestry, who was everything most men would want. She did all the right things except for marrying a man who did not know how to be married. We had met in Seattle, drawn by our contrasts, and we exchanged vows without really revealing our depths, and this last part was more true of me. I was still gauging those depths, still feeling out the contours, still grasping for the right words to match. I brought a certain melancholy into the marriage, and I grew disappointed when the marriage did not fix me. Then the most craven thing: I secretly blamed her for my unhappiness and withdrew without saying why.

Marriage stories are always evolving. And so I say this knowing that it's an idea still undergoing mutation: I don't believe our ethnic and racial backgrounds played a huge role in our breaking apart, but they may have played *a* role. I have friends who tell me now that they played a bigger role than I've ever been willing to acknowledge. One of those friends, Lizbeth, who met my wife and her family at our wedding, told me after the split that she knew within moments the marriage would not last.

"And you didn't say anything," I said.

"Please," she said. "You weren't in a position to hear."

"Very true. I probably would've resented you—"

Lizbeth nodded. "Probably."

"Okay, well, you've said it now. Are you going to tell me what you saw?"

Lizbeth thought a moment. She looked at me with a mix of apology and sympathy. "Your 'hood and her 'hood are on opposite ends of town, and never the twain shall meet," she said.

I grew irritated. "Which means what, exactly?"

Lizbeth cocked her head, meeting my irritation with her own. Her expression said *Come on. Don't be naïve.* Lizbeth's father was black, her mother white. Lizbeth identified as black, and sometimes she saw things in black and white. Not always. Sometimes. This may have been one of those times. For whatever reason, I didn't pursue the conversation, and I regret it because, as I think of it, she could have meant a lot of things. After all, it was our difference that had attracted my wife and me to each other at the start.

I was a child of Asia, the color of earth; my wife was descended from Europe, the soft hue of porcelain. I was shorter, she was taller ("but the same height in bed," we used to tell each other). I was mercurial, she was steady. I tended to live in my head, she through her body. My family was emotional, demonstrative, and given to drama; hers was circumspect, formal, and prone to keeping its dramas private.

Our families had nothing to say to each other beyond the niceties, which were usually covered within twenty or thirty seconds. And neither was interested in cultivating deeper engagement, almost as if they knew the futility of it. My wife and I underestimated the impact of our families'—and perhaps our ancestors'—reach into our souls. We thought ourselves above all that, a modern couple born anew in a new land, committed to the idea that love transcended all. I still believe love *can* transcend all, but ours didn't. It's possible that we each inherited something of the molecular makeup of our clans, which we then carried unknowingly into the marriage, and which, after the initial euphoria waned, resulted in a profound chasm that neither of us could bridge. It's possible.

I remember signing the divorce papers at a coffee shop on Greenwood Avenue, my chest heaving with sobs, overcome by the fact of my utter failure. My soon-to-be ex-wife, still lovely, deserving better, watched me calmly from across the table, holding herself together with a dignity that I could not comprehend.

I vowed never to marry again.

A few years later, I married again.

Melissa and I have now been together, counting our courtship, for twenty years. She is smart and beautiful, and better than me in most ways that matter: kinder, more patient, more giving, and though gentle in spirit, far tougher than I. She puts up with my dark moods partly by not being absorbed by them. She also doesn't let me get too far lost in my reverie before figuring out a way to bring me back. By being a patient and dedicated mother to our two daughters—the older one from my first marriage—and to the nieces and nephews who are like our own children, she teaches me to be a far better father than I was likely to be. People who've known me a long time, and who aren't afraid to be blunt, like my friend Lizbeth, have confided that they saw our good fit almost immediately. It was a gut observation. Their gut corroborated my gut.

It wasn't just that Melissa was also a child of Filipino immigrants, but this played a role, no question. Our skin color and features and mannerisms identified us as members of the same tribe. Our childhoods in America were similar enough that we did not need to explain them to each other. We knew each other's backstory. But beyond the common citizenship of our otherness, we shared a similar manner of engaging life, as born observers who grappled with words and ideas. We met in a newsroom, two journalists curious about the world and trying to make sense of it. Our muses aligned.

I experienced a relaxed way of being that I had never known before. Melissa has told me that she felt the same natural fit from the start. She'd had five previous boyfriends before me, all white. Being with me, she said, brought a sense of coming home. The home was messy, in need of constant repair, and had too dark and labyrinthine a basement with too many spiders, but still it was a place where her deepest self could relax.

Regarding this investigation, this secret education of mine, she accepts that I'm driven to pursue it even if she doesn't completely understand the urgency that drives me. She doesn't see me as *more deficient* than other men, although she has told me in so many ways that she understands how circumstances could create the feeling in me. She gets it. She also stays detached from it. She supports my investigation the way some wives put up with husbands who tinker night after night in the garage on a project vitally important only to the tinkerer. The end product will be what it will be. It's the tinkering, the quieting of some inner noise, that matters. She keeps a watchful eye, and she's witnessed changes in me.

Starting in the mid-eighties and continuing through the nineties, I felt increasingly that my inferiority story, my tale of Asian woe, was in need of major overhaul. The mythology that helped me understand my early experiences in the land of the giants seemed less and less convincing as I witnessed the enlarging presence of Asian influence everywhere I traveled, and I did not have to venture far.

It wasn't just the Asian restaurants and stores, the yoga and meditation and martial arts studios, the neighborhoods and shopping centers with Asian names spelled out in Asian characters, the Hondas and Toyotas and Nissans on the freeways, and the Sonys and Panasonics and Samsungs on store shelves in every city in America.

It wasn't just the steady drumbeat of news of Asia's economic rise, one miracle after another, first Japan, then China, then India and the smaller "tiger" nations like Taiwan, Singapore, Hong Kong, South Korea, and Malaysia. Their collective rise may be the story of the century. During the Western Industrial Revolution, standards of living rose perhaps 50 percent in a single lifespan—unprecedented in human history. At current rates, standards of living in Asia may rise, in one lifespan, 10,000 percent. By 2020, according to projections, three of the four largest economies in the world will be on the Asian continent.

The sweeping pronouncements coincided with the rise of Asians here in the United States. Asian Americans, relative newcomers to the New World, were dubbed the "model minority," with higher-than-average education and earnings, and with a faster-than-typical climb up the socioeconomic ladder. Asian women made a strong impression in the Western imagination. Asian men began to impose their presence in public life. The latter was driven home to me in a back-to-back encounter in 1998.

That year I traveled to Honolulu on a fellowship and managed to gain, with other fellows, an audience with Hawaii's governor, Ben Cayetano. I hadn't known about him before then. We met in his executive chamber at the state capitol. He was broadly built with thick black hair and kindly eyes that were well on their way to becoming haggard. For twenty minutes he answered our questions with businesslike aplomb. I observed more than I listened. I marveled at the sight of him. When we shook hands at the end, I noted that his skin was the exact—*exact*—shade of earth as mine. I saw the outlines of my own face in his.

Soon after, at a public forum in Seattle, I met the governor of Washington, Gary Locke, with whom I'd had professional contact ever since he was a lowly county executive. Locke's was a quintessential American story. Born in a public housing project and raised by Chinese-speaking parents, he did not utter his first English words until kindergarten. Now he was the guy in charge of my home state, the boss of everyone at the forum, and he exuded the appropriate authority, subtly but unmistakably.

So, within a few weeks, I had come face-to-face with two articulate, firm-fisted, formidable chief executives who happened to be Asian men. In fact, they were the only two Asian American governors in the country. Cayetano had sprung from the land of Lapu Lapu in the Philippines; Locke was a descendant of the Middle Kingdom. I remember thinking that life was trying to rearrange the old storylines in my head. To redraw the pictures and give me new characters to dwell on.

At about the time Locke's first term ended, a dynamo from Japan hit Seattle. His name was Ichiro Suzuki, a slender, grinning, sunglass-wearing baseball player with a rocket arm who in a short burst of time became the most popular athlete in the Pacific Northwest. He was no mixed-blood *nisei* with distant Asian roots but was Japan-born, bred, fed, and cultivated, a product of the Nippon Professional Baseball league. He was listed as five foot eleven but was probably five foot nine and 170 pounds, if you counted cleats, helmet, and bat. The big question was, how would he fare in the land of the giants?

He answered emphatically. In his first season with the Mariners, he earned Rookie of the Year and Most Valuable Player honors—a rare feat in the same year—and led the team to 116 wins, tying the all-time record for the modern era. He became Seattle's own rock star, a single-name phenomenon like Madonna and Bono. *Ichiro!* Paparazzi stalked him. Autograph-seekers mobbed him. His face got plastered on buildings and billboards, posters and T-shirts. Life-size Ichiro cutouts accosted me in every corridor. Ichiro bobbleheads bobbed on every desk. The adoration seemed universal. Ichiro could barely say "thank you" in English.

I recall a moment walking out of a theater in the University District after watching *Matrix Reloaded* with a friend. She was a fan of Keanu Reeves. "He's part Hawaiian and Chinese, you know," I informed her.

"I knew he was part something," she said.

"The one-drop rule says he's an Asian guy."

"You Asian guys are taking over."

My friend punched me in the arm, and I chuckled. But seriously. The man who ran the state was Chinese. The most popular jock in the region was Japanese. Two recent Pulitzer winners at the *Seattle Times* were Filipinos, one, Byron Acohido, with a trace of Korean; the other guy was me. My friend and I had both driven to the theater in Toyotas, and we had just walked out of a blockbuster movie whose male lead, a rising cult star at the time, was part Chinese and Hawaiian. The developments that made for such a crystallizing moment weren't isolated to Seattle or the West Coast. What I was witnessing at home was happening in varying degrees all over the country.

All these developments qualified as corroborating evidence in my personal investigation. I came to realize something about the nature of my inquiry: it was almost as if I had put my own manhood on trial, and I was playing both prosecutor and defense. The prosecutor in me, the voice of self-condemnation, was relentless and vicious, and talented at using all manner of evidence. He had no shortage of material, and he wanted to keep me in that hole of inferiority forever. The defense used all the good stuff happening around me, particularly the increasing public profile of Asian men, as part of its case for lifting me up: worthiness by association.

No question these external developments were necessary to the case. They symbolized the collapse of old barriers, and expanded the silhouette of the Asian man in the cultural imagination. I felt encouraged and reassured, and experienced moments of visceral inspiration. Watching Liu Xiang cross the finish line *still* quietly exhilarates me (and now I can watch him on YouTube). Seeing Cayetano and Locke in the flesh, and watching Ichiro do his thing at Safeco Field, instilled a kind of civic pride. I could begin to identify myself as part of a recognizable group making strides in the public arena. These men who broke out into the open and made themselves seen made *me* feel more seen.

But looking back on this period in my life when I was starting to rise from the hole, I realize that something else lifted me up in an altogether different way, something from which I did not expect spiritual uplift. This something else created an *inner* substance and cohesiveness that gave me the sense of being weightier, more solid, more able to push off the ground.

What gave me this substance was work. Specifically, doing meaningful work that allowed me to use the limited gifts I was born with. Work that allowed me to immerse myself in issues of vital interest to me. In the best moments, work that made a difference to someone or something other than myself.

The curious thing was that, for a long time, I regarded my

decision to become a journalist as a purely practical measure — a way to pay rent and start chipping away at my student loans. In college, I had learned that I was a decent writer. Writing was really the only thing I seemed proficient at, although it always came harder for me than for others, for whom writing seemed to come as naturally as exhaling. I knew an accomplished writer who used to whistle in birdsong as he typed. For me, it was always hard labor involving quiet despair. Sometimes weeping occurred. But I always got it done and would earn mostly favorable assessments. So after a series of get-by jobs — shoe salesman, fish slimer, credit analyst, janitor — I decided to try my hand at writing for a newspaper.

The idea was to work as a reporter for a couple of years, hone my skills, and then do something else, something grander: go into law and represent the underrepresented, like Atticus Finch in *To Kill a Mockingbird.* Or become a writing monk like Thomas Merton or a typing hobo like Jack Kerouac.

I had no abiding interest in news, that is, events unfolding right now. I never cared about getting a story first, or covering the latest mishap or disaster. But I resigned myself to having to pay my dues by covering the *Sturm und Drang* of routine events before being granted the privilege to work on more ambitious pieces. In short order I began writing features, and my editors found I had a knack for the in-depth stuff, what newspapers call "enterprise." Small successes led to larger ones, which led to opportunities at bigger and better newspapers. As so often happens, what was supposed to be a temporary stint turned into a career. It reminds me of the Gary Snyder poem "Hay for the Horses," about an old guy who started bucking hay at seventeen. The poem ends, *I thought, that day I started, / I sure would hate to do this all my life. / And dammit, that's just what / I've gone and done.* I kept that poem tacked on my wall for years as a reminder, and a cautionary tale, to myself. But I went and did it anyway. I worked in newspapers for more than two decades.

Journalism was never a form-fitting vocation for me. It never clung skintight over my soul the way it seemed to do for some. I never felt completely at home in the newsroom; my story proposals were often seen as too abstract, my work rate too slow, my competitive drive too lax. But the job was a good enough fit on enough days to keep me from quitting, and to keep the bosses from firing me. It provided a steady paycheck and an occasional moment of creative euphoria.

I lived for those small moments. Whenever I focused my attention on topics important to me, I produced stories that elicited stronger reaction. The more I engaged my insides, the better my work. The cause and effect seem so obvious now, constituting such a basic truth, but it took a shameful number of years for me to connect those dots.

"Don't you get tired of writing about the same thing all the time?" a colleague once asked me. This colleague and I were roughly the same age and rank in the newsroom. He was ultra-competitive, and he saw me as a rival even though we were on the same team. I never understood it. He made it known that he thought of me as a one-trick pony. His question was a way of putting me down.

By "the same thing" he meant race.

It's true that many of my stories involved racial and ethnic minorities. One of my very first beats was covering street gangs, which had become a plague in Seattle beginning in the late 1980s. When I first started, the problem involved mostly black kids, and I wrote many stories with African Americans as the main characters. By the time I left the beat years later, the most violent gang crimes were being committed by young Asians and Pacific Islanders—primarily Filipino, Vietnamese, Cambodian, and Samoan. Some Tongans and Fijians fell into the mix, too. I wrote an untold number of stories through the 1990s and early 2000s exploring these communities. In doing so, I saw the seedy underside of the so-called model minority. I got exposed to the darker winds of the yellow tornado.

While immersed in a story about Cambodians, I met a woman in her thirties who lived in a local housing project. Her white and black neighbors, and even some of her Asian neighbors, thought her too detached and strange as she walked to and from the bus stop, sometimes talking to herself. Her name was Arun. This is the story that she and her therapist would come to tell me:

One day outside Phnom Penh, when she was a young girl, Khmer Rouge soldiers took her family to a field. They ordered them to dig a hole and line up at the edge. A soldier shot several family members in the head. When the soldier ran out of bullets, he hit the others with his rifle butt and stabbed Arun in the chest. The crumpled bodies were pushed into the hole, and the hole was covered with dirt. Arun survived, crawled out, and walked for days before finally crossing into Thailand. At the border, she was raped by soldiers. She was placed in a series of refugee camps, where she lived among strangers for several years before somehow landing in America—first in Virginia, then California, and now Washington. Arun lived alone with her memories. She worked in a nail salon.

I wrote about her and other Cambodians who had survived Pol Pot's genocide. They were walking wounded, and their wounds were not so easily detected. Many lived with trauma that never healed, in some cases, not even recognized by themselves. They were simply *chakourt*. Crazy. And they were Americans now, citizens who had endured unimaginable horrors, like the Jews of the Nazi Holocaust, but whose plight was barely acknowledged by the society they now called home.

For many years I wrote about Native Americans, with whom I deeply identified. I was often mistaken for Native. Really, from a certain standpoint, we were one and the same, Native Americans being the descendants of Asian people who crossed the Bering Strait, in geological time, not that long ago.

At its closest point, only fifty-five miles separate Alaska from Siberia, and the Asian connection remains vital to thousands of Natives in Alaska's far west. In 2004, I traveled to the Yukon-Kuskokwim Delta to write about a Yupik Eskimo village on the verge of disappearing under rising waters. The Yupik people, who rely on hunting and fishing for subsistence, originated in Siberia. Alaskan Yupiks maintain close cultural ties with their relatives on the other side of the Bering Sea. In their daily lives, Alaskan Yupiks have more in common with Siberian Yupiks than with the vast majority of fellow Americans in the Lower Forty-eight.

I devoured Native American stories as if they were my own, and I took it personally as I read accounts of what happened to Native people when Europeans arrived.

I stood in the exact spot at Wounded Knee Creek on the Pine Ridge Indian Reservation where Chief Spotted Elk (known as Big Foot among U.S. soldiers) and more than two hundred children, women, and men of the Sioux nation were gunned down by the Seventh Cavalry. The photographs of the massacre reminded me of the images of My Lai which made such an impression on me as a boy.

I sat at the gravesite of Chief Joseph, the great Nez Perce leader whose people were defeated, the survivors forced onto reservations far removed from their traditional grounds. I broke bread with a living descendant of Chief Joseph, a cranky pony-tailed man named Taz Conner, who lost both feet to diabetes and got around in a listing wheelchair. He spoke of his desire to return to his tribe's ancestral lands, the land of winding waters, as the Nez Perce called it.

"I don't know if I'll see it in my life," Conner told me. "I'm supposed to be dead three times now. Diabetes. Kidney failure. All sorts of complications. The first time, they told me I had ten days to live; the next time, one month. Now they're saying, 'How come you're still alive?' If it's going to happen in my lifetime, it better happen pretty soon."

So I guess you could say I've written a lot about one thing as a journalist. But I hardly ever saw it as exclusively about race. To my mind, it was more about telling stories of people who existed outside the mainstream's field of vision. Invisible people. Barely discernible beings who lived among us, sometimes right next door, or pushing a cart down the same cereal aisle at Safeway, who moved through life largely unseen because their stories were largely untold. Or if they had been told, they were told as exotic tales of outsiders, *them* as opposed to *us,* or as history rather than living narratives playing out here and now.

My father lived with this. No one in America ever knew or cared about his life prior to arriving stateside, and even his wife and children got too busy to pay attention to him, to pay homage to his experiences.

A glimpse: Six of his seven siblings died in infancy from illnesses and accidents. His father drifted away and his mother lost her mind. She was killed when a church van she was riding in drove off a cliff. My father practically raised himself. During the Japanese invasion of the islands, he joined a band of guerrillas and almost starved to death in the mountains of Mindanao. A lot of people he knew simply disappeared. He got malaria and was nursed back to health by a young woman who was later raped in his presence, and for the rest of his life he hated himself for not protecting her. He was fourteen at the time. He had nightmares about it until he was an old man. In America my father made a few friends, but none of them really knew him because they didn't know his story.

Here's what I'm getting at. My own lifelong sense of feeling invisible, and living with others like my father who experienced the same, somehow became useful. I developed the sensory apparatus to apprehend fellow invisibles. In making them seen, even for just a few column inches on a Sunday morning, I had found a purpose. Not the grand one that I fantasized as a young man, but a workaday purpose. A blue-collar, punch-the-clock purpose. Work gave me a place to go every day, gave me something outside myself to care about. As an unintended consequence, an accident, I began developing this inner sense of ballast that my father never found in the land of the giants, and I considered myself fortunate.

One other thing. I discovered that I had unwittingly adopted a new identity. I began to see myself as a chronicler, a dude who, instead of running away from calamity ran toward it, got as close to the center as possible, took notes, and then recounted the story. The realization that I had taken on a new way of viewing myself helped me to understand a Walt Whitman line that had always intrigued me: *I am large, I contain multitudes.* Whether we acknowledge it or not, we are each a plethora of identities. I was an Asian, yes. I was also a journalist. A husband. A son. A brother. A friend. An agnostic who still prayed to Jesus. A part-time misanthrope. A crank, a wastrel, a sap. An islander, an immigrant, a citizen of the United States, an inhabitant of planet Earth.

Acknowledging that I contained multitudes meant that I did not have to place the burden of my worth on any one of my identities. It was like growing additional legs; one leg could collapse and I could keep moving. In this way I got to be more sure-footed, less vulnerable to being knocked back down to the bottom of the hole. I had already spent too much time down there. I had made some progress, but there was still so much distance to cover. At some point, the opening at the top no longer seemed so unreachable. I found toeholds and began to rise.

12

"What Men Are Supposed to Do"

A strange old man
Stops me,
Looking out of my deep mirror.

—Hitomaro

My father died apologizing. He was sorry for everything. He still had so much to do, so much to prove. He had scraps of paper, with lists and plans for businesses he wanted to start, tucked away in drawers, books, briefcases that he hadn't opened in ages. If only his body could hold it together for five more years, he told me. Just five. He would change things. He would stop wasting so much time on trivialities. He would focus, and behave, and pray night and day, and maybe God would finally reward him with the big break he'd dreamed about since he was a young man who didn't know better. A young man who believed that breaks and good fortune came only to those who deserved them.

I had heard the litany before. His contrition in those last months, when he diminished day by day into a trembling branch, was just a concentrated version of sentiments he'd shared with me for many years: he wasn't rich enough, educated enough, fluent enough, Americanized enough. He didn't provide enough. He wasn't connected enough, mindful enough, moral enough, disciplined enough, persistent enough, powerful enough. *Man enough.* He didn't do what men are supposed to do.

He never seemed to hear me when I told him none of that counted in the end. He would look at me searchingly, appearing for a moment to consider my words. Then the moment would pass. "You don't understand, *anak,*" he'd say. His mind was made up. It had been made up for a long time. He was someone who had chosen a narrative about his life and remained loyal to it no matter what, even if it brought him endless misery. He had a gift for misery, a gift he inherited and passed on.

The thing was, when I reassured him, I mostly meant it, and I meant it more and more in the last decade of his life as his health deteriorated and my views broadened. My definitions of manhood, once aligned with his, had been undergoing great change—partly out of will, and partly because life had dropped individuals in my path who expanded the outlines of my thinking. As my thinking changed, so did the way I viewed my father.

One who made an impression was a man I met through work. Tom Nakao was a *yonsei,* a third-generation Japanese American in his forties. I had been researching a story about a gang murder of a teenage girl named Melissa Fernandes, and Tom was helping police with the investigation. I remember meeting him at a pool hall on Beacon Hill, a hangout for one of the most violent gangs in the city.

I arrived first and took a seat in the back. It was a one-room joint with half its fluorescent bulbs burned out and two ratty pool tables. A low ceiling gave the place a bear-den feel. The room smelled of old vinyl, smoke, and Lysol. Crushed beer cans spilled from bulging plastic liners next to my table. Eight or nine guys were shooting pool, a couple of them giving me hard stares between shots. One flicked a cigarette butt just beneath my chair.

When Tom walked in, they dropped everything and approached him, hands outstretched. He made contact with each, exchanging bits of news, and wrapping a couple of them in an ursine embrace.

He was a big man, tall with broad shoulders and a paunch that stretched taut his threadbare jacket. Long black hair framed a face with fleshy cheeks. Wire-rimmed glasses perched on an aristocratic nose. He walked over to me, put a meaty hand on my shoulder, and then told the young men who were all looking at us, "This guy's okay. Respect." A moment later, one of the men who had been staring daggers offered me a beer.

Tom was a "gang interventionist," a title he never cared for. He saw himself more as a surrogate father to young men whose real fathers were dead, absent, or abusive. By getting involved in the young men's lives, he made inroads into the local Asian immigrant communities, which others found impenetrable. His job was to intervene on behalf of schools, but the police increasingly tapped him for help. After the Fernandes shooting, he was summoned almost immediately. He made some calls, hit the known hangouts, and within hours learned the identities of the young Asian men involved in the murder.

For ninety minutes at the pool hall, Tom gave me the inside details of the shooting. It was the first of several meetings between us. He was to remain primarily an anonymous source. The shooter, he told me, was a sixteen-year-old gangster wannabe who was showing off to his friends and trying to scare a rival gang member in the crowd. The kid didn't know how to handle a Mac 10, and he sprayed bullets into the air, one of them hitting Melissa Fernandes.

"He's got to pay the price now," he said. "Consequences. I'm always telling them about consequences. Now they know what I'm talking about. Now they're sorry, but sorry doesn't bring back Melissa. Thing is, these kids aren't evil. Most of them are good kids. Smart kids. They're just caught up in this gangster game. We need to show them a different game."

Tom said he'd been caught up in the same scene growing up on Cleveland's East Side. A former cop in the neighborhood took an interest in him, guided him off the streets. Eventually Tom went to school, got married, and spent the better part of two decades as a businessman in the Midwest. A few of his investments did well enough for him to go into modest retirement. He moved to Seattle and, recalling the cop who'd helped him, decided to spend his days showing young men a different way to carry themselves.

Over the next several years — in dimly lit rooms, playgrounds, back alleys, and pool halls — Tom introduced me to other local men who were similarly called. Tim Cordova, an innovative teacher and writer who worked primarily with young Filipinos. Fia Faletogo, a hulking California transplant who spent his days mentoring, and sometimes collaring, young Samoans, Hawaiians, and Tongans. Ron Carr, a former prizefighter turned social worker from Chicago who used boxing to instill discipline in aimless young black, Vietnamese, and Cambodian boys. Winslow Khamkeo, an outreach worker who targeted young Southeast Asian refugees living perilously on the edge, putting himself on that edge to maintain contact.

These men worked in battle zones, and a few of them had the bullet and knife scars to prove it. They did not earn much money and never received public recognition. Anyone who spent time with them realized they were driven by a different set of imperatives. They weren't saints, but they had answered a call beyond the usual climb up the ladder of status and self-aggrandizement, spending the majority of their hours looking after the welfare of people who needed looking after. And they had figured out how to enact their compassion in a distinctly masculine way: venturing into dangerous territory, exposing themselves to high risk, and facing the unknown in the manner of the archetypal explorers and conquerors.

It's mystifying why, in my explorations of manhood, it took me so long to acknowledge men who inhabited a more expansive moral horizon. I'm pretty sure I understood the idea of compassion even as a boy. I knew that it was extolled by the likes of Buddha, Muhammad, and Jesus — in some ways the ultimate men, but men in the abstract, closer to divine than I could hope to be. No man close to me had exhibited this virtue as a daily reality, a way of life that I could, or should, strive for.

Perhaps such men were around during my first decades, only I didn't see them, preoccupied as I was with the frenzy of fitting in. But as I got older and calmer, my life stabilized enough for me to take note of subtler realities, including the outlines of a masculine ideal shaped by the personal desire to do good. What a radical concept to discover in adulthood: doing good as a hallmark of a real man.

The vision continued to expand. Tom functioned in one

of the less visible swirls of the yellow tornado. He exerted his power in the shadowy parts, and in ways not meant to be seen by the public. I've come to appreciate other, more visible manifestations of compassionate power. Giving can manifest as a giving *in* to a creative muse that transports people beyond the pedestrian. Musicians, artists, craftsmen, designers, storytellers of every medium, and prodigies in science and technology belong in this category. There's the giving in to religious faith, which can produce men and women who forsake everything to care for the suffering and abject. There's also the giving in to conscience.

In 2007 I wrote a story about a baby-faced army lieutenant named Ehren Watada, the first commissioned officer in the United States to publicly refuse deployment to Iraq. "If our country needed defending, I'd be the first one to pick up a rifle," he told me. "But I won't be part of a war that I believe is criminal." Watada, an Asian American from Hawaii, requested deployment to Afghanistan instead. The army decided to court-martial him. I sat in Watada's cramped living room as he packed up his gear, and I wondered whether I would have heeded my conscience as he had.

There's the category of givers who write checks. A 2013 front-page story in the *New York Times* described a new affluent class of Asian Americans donating large sums to prestigious schools. Examples: New York philanthropists Anthony Wang and his wife, Lulu, donated $25 million to Wellesley College. Donor Oscar Tan gave $25 million to Phillips Academy Andover. Jerry Yang, co-founder of Yahoo! Inc., gave $75 million to Stanford University, one of several donations he's made to his alma mater. It seemed to signal a new chapter in the Asian American story.

Asian forebears had crossed the ocean, survived, and thrived, and now their progeny were giving to causes that enabled others to cross their own frontiers.

It all went into a new and still evolving equation in my head. Ehren Watada, Tom Nakao, and Jerry Yang expanded the perimeter of my definition of manliness. Maybe I'll never need to stand up to the United States government, but perhaps I could stand up *for* something. Maybe I could never be as emotionally available and humble as Tom Nakao, nor as epically generous and recognized as Jerry Yang, but I could be my own kind of giver somewhere on the continuum in between them.

My father gave me more than he could know.

For a lot of years after I left home, I sought to be as different from him as possible. Whenever I determined that he'd gone in one direction, I resolved to go in the other, and to travel as far as I could so that I could someday confront him with my utter oppositeness. But in searching out what kind of man I wanted to be, what kind of man I *could* be, I have lived out T. S. Eliot's line about ending my explorations by arriving where I began.

I have become a lot like my father. Biology contributed to it. His disposition and mine are identical, our talent for melancholy hardwired and set. Our egos match. Our vanities correspond. I'm drawn to the same vices, among them the occasional and unpredictable urge to carouse and philander. I share some of his underlying tenderness, and the secret embarrassment about it. My hands resemble his, right down to the pattern of veins. I perceive the lines of my face in photographs of him. In home videos, my mannerisms come across as cheap imitations of his. In my way of speaking, even of pausing, I hear him. We have the same crooked smile, and laugh at the same things. We fake laughter at the same kinds of unfunny jokes.

I want to think that I could be as courageous as him. Not just brave in the moment, as in pushing back when someone pushes you. But deeply and solemnly courageous, as he was in leaving everything familiar behind in midlife and starting over in a strange new land where he would not have the years necessary to become fluent enough, or culturally adept enough, ever to feel truly capable again. He took the risk so his children could experience that fluency, could feel that capability and promise.

I want to think I could be as playful as he was. My siblings and I remember those early Sunday mornings when he would tickle and kiss us awake, acting silly and bellowing in his cartoonish baritone. I act silly in the same way. I can reach the same low notes.

I know I can dream like him. He was a big-time dreamer. It was *his* dream that transported us to America. My mother used to say that dreaming was what he did best. When she was mad at him, which was often, she'd say it was *all* he did. He never seemed to understand that most of us have to adjust our dreams continually and let go of the most fanciful of them eventually. I have to remind myself of this. The letting-go part I have a hard time with too.

I'd like to think that I could be as generous, although not to the crazy degree that he sometimes was, giving away not only his last dollar but his wife's and children's as well. He would have given away our house if someone had asked. He seemed oblivious to consequences. It was that pattern again: he held things tight, then like a burst dam, he gave everything. This applied not only to money and possessions but to his insides, too. When someone inquired, he spilled all. He would stun people with his capacity to be mercilessly candid. He could not seem to help it. Too often, I can't either.

He was generous in one way that I *do* try to emulate. Some of my fondest memories are of times when we did distinctly unimportant things together, like picking blackberries or walking ten blocks to the hardware store or staying up late tying fishing hooks, and we did these things while feeling quietly connected. Our fishing trips were like that. Those days when we did not catch anything were almost as good as the days when we did. He made it a point to have these moments with me, and I try to have them with my daughters.

One daughter is in college, the other in middle school. On a recent typical day, the older one and I went to a Euro-Japanese café down the street and talked about her sociology class over tempura udon. She's learning about "gender construction in the media," and fortunately, I knew what she was talking about.

A little while later, the younger daughter and I went to Baskin-Robbins for cotton candy ice cream, and then we took the dog to a vacant field to play fetch for a half hour. In between tosses, we talked about *The Walking Dead,* a show we weirdly both like. The field is just behind the dance studio where they both belong to competitive hip-hop teams. Late in the afternoon, I drove them to dance practice. We talked in the car about an upcoming competition in Vancouver. They were both really excited about it, and I just listened to them. Later, over dinner, we talked about the logistics of the Vancouver trip, and they showed me a couple of new moves in their dance numbers. I showed them some of my dance moves from the eighties, which made them wince. It also made them laugh. That's it. That's all we did, and it was a good day.

I can't afford to take my girls on many exotic vacations or

shower them with a whole lot of expensive gifts. I don't have any spectacular talents they can brag about to their friends. I work too much, detach too often, and can sustain a cranky mood for much too long. But despite all of this, my girls seem to like my company. They're at ease with me. They laugh a lot, cry some, whine some, and tell me ridiculous stories that no one else would find interesting. We can and do talk about anything. They can be utterly themselves. My gut tells me that they find me acceptable. Occasionally, they say they love me.

I felt the same about my father.

The other day I opened his urn, which sits on a cabinet in my dining room. I see it every day. It was meant to be temporary. The plan was to spread his ashes in the village in Mindanao where he was born as soon as the region stabilizes (there's been a violent and long-running insurgency). Inside the urn is a clear zip-lock bag containing my father's ashes, slightly larger than the ones used for sandwiches. I opened the bag, and a puff of gray dust escaped. Maybe someday my own ashes will sit on a mantel, and my daughters will occasionally stop to gaze at everything that I was. The thought might occur to them, as it did to me, that no matter what we do or don't do in life, we all end up insubstantial enough to fit in a sandwich bag.

My father thought he had failed as a man. He would not accept what I'm trying to teach myself to accept: that he was just a man, the same as most other men. Fearful. Vain. Deeply flawed. Constantly wanting. Chronically anxious. Born on the outside edge of the Garden, and living always with a suspicion of unworthiness.

I've thought about him more often than I did when he was alive. I've thought about our last conversations in those months before his presence diminished to mere breath and bone, and he could no longer hear me. I wish I had told him that the circumstances of his life and much of what he felt were not entirely his fault. *Bahala na. Mahal kita.* He did the best he could. In the ways that really mattered, he was enough.

13

"One of Us, Not One of Us"

Before I can come to the end
Of an endless tale, the children
Have brought out the wine.

— Tu Fu

W ho is that guy?" I overheard a young woman in a white Nike cap say to her companion, a pale young man who was wearing an identical cap. Standing directly behind them, I had been watching their matching caps swivel in unison for a half hour as they watched the action move from one end of the court to the other. "Honey, did you hear me? Who is he?" she said. He watched the end of the play before answering.

"I don't remember his name, but I've played with him," he said. "I know. He's good. Korean, I think."

The couple was part of a small crowd that had gathered to watch a pickup basketball game at the rec center on the campus of the University of Oregon, where I now teach. It was a typical weeknight game, with a mix of high school has-beens and jock wannabes huffing up and down the court, lobbing Hail Mary shots. A few of them had skill. The one standout was Aaron Lee, about whom people in the crowd whispered.

"Watch this, watch this . . . ," a young man on the sidelines said to his friends, who leaned forward to focus on the action. The ball was passed to Aaron at the top of the key. In what seemed like a single fluid motion, he caught the ball, moved past his defender, took two steps, and then rose in the air for an effortless slam dunk.

The defender laughed and clapped his hands in his own small tribute. "Okay," the young Nike-capped woman said to her Nike-capped companion, who stood and gave a loud whistle. "Bet you've never seen that before," he said. The young guys on the sidelines cheered. "Stop showing off, Aaron!" someone yelled from the back of the gym.

Aaron flashed an easy smile and trotted up the court. Just another play, another night at the rec center. Every Monday evening, eight thirty, he was here. Unless there was someone from the UO basketball team there, Aaron was usually the best player in the house. He was tall, leanly muscled. As a junior he walked on and stayed on as a wide receiver on the Oregon Ducks football team, one of the top teams in the country. Like so many young Asian American men who grew up on top sirloin and russets — in his case, along with kimchi and rice — Aaron towered over his smallish immigrant parents.

"People don't think Asians can play," he said to me later. "I like proving them wrong. *Give me the ball.*"

As I got to know him, I found that the attitude extended beyond the basketball court and the football field. It was apparent in the way he carried himself off the field — not with a swagger, but with an air of understated, almost gentle confidence. A soft glint of optimism shone from his eyes even when he thought no one was looking.

Still, he was young and harbored the same anxiety about the future that any twenty-year-old would who hadn't yet decided on a vocation, or even a major. He had narrowed it down to either business or education, and this week he was leaning toward education. "Teaching elementary school would be fun," he told me. "I enjoy younger kids. They're so open to you. I think I might be a good teacher." His anxiety surfaced only in moments. What emanated from him most of the time was an aura of calm, as if he knew deep down that whichever path he chose would work out. He would be fine. Where this outlook came from, Lee couldn't say, and I could only guess: family, faith, genetics, good fortune, a synergistic blending of all.

His large extended family attended the same Korean church in the same leafy suburb. His mother ran a day care center and enjoyed it. His father was a longtime postal worker. The kids were all scholar-athletes who, it was presumed without question, would finish college and becoming contributing citizens. Solid, one and all. The Lees had stakes dug deep in their community. It kept occurring to me that at least part of Aaron's optimism was grounded in the awareness, not entirely conscious, that he had a place in the culture that his family now called home.

Aaron Lees — self-possessed, socially promising, physically substantial young Asian standouts — would have been difficult to find in the America that my family entered in the 1960s, when Asians numbered fewer than a million and were scattered far and wide. There may have been some, but I did not meet any as we hopscotched from coast to coast. The number of Asian

Americans today is 18 million (with 5 million more in Canada), a third of them born in North America. It says a lot about the changes that have transpired since my family's arrival that you can now find Aaron Lees at any major university in the country, and you wouldn't have to look that hard.

The University of Oregon is in one way a microcosm of the country. Of the 25,000 students, roughly 5 percent—same as the national figure—are Asian American. You can spot them anywhere and everywhere on campus. Strolling down Thirteenth Avenue, which cuts through the heart of the university, you would notice an easy intermixing of genders and races and ethnicities. Among the combinations, you'd see Asian women walking with men of different hues, and Asian men side by side with white, black, brown, and yellow women of all shapes and eccentricities.

In the past several years, I've noticed many more Asian men with Asian women. I could not say for sure whether this is a new social trend, but recent studies indicate that it might be the beginning of one. Since about 2010, researchers have found that Asians in the United States are increasingly dating and marrying other Asians. Perhaps the Asian population has grown substantially enough that there are now many more eligible men and women exposed to one another. The choices and opportunities have expanded. I'd also like to think the trend signals, in part, an expanded sense of self-acceptance among the younger generations. Maybe if Leny and I met up in college today, we'd have a chance.

Outside Prince Lucien Campbell Hall on Tuesday and Thursday mornings, you'd find Josh Volvovic—sweatpants, white T-shirt, black beanie, earbuds—locking up his bike before heading to class in the tallest building on campus. He is Filipino American, a senior with a double major in political science and journalism, a Phi Beta Kappa with a 3.75 grade point average. He sports a sturdy weight-room build. His large, expressive eyes and thousand-watt smile served him well over the summer as an intern at NBC's *Today Show* in New York. He went to work every weekday morning at Rockefeller Center. *The Rock!* NBC wanted him back after graduation, but Josh decided on law school instead. He applied to three, got accepted to all. I was one of his recommenders.

Sometimes I would look at him and think to myself: *limitless.* The guy was a new-era Asian American. He could go anywhere, do anything. The world was his flower to pluck. As if reading my mind one day recently, he told me, "I know there are possibilities, but I tell myself I have to be reasonable with my expectations." He might have been suggesting that I be reasonable about mine. It was typical of the modest young man I'd come to know. He was twenty-one, still wary of acknowledging the massive potential that other people saw in him.

On Monday and Wednesday mornings, Josh biked across campus to Allen Hall, where Prapat Nujoy studied public relations. I did not know many seniors on campus with a more kaleidoscopic background than Prapat: half Thai, half Chinese; born in humid Houston of immigrant parents; played high school football in Alaska's frigid Kenai Peninsula; earned a bachelor's in music performance from the University of Alaska and was now working on his second bachelor's at UO. He played bass trombone for Oregon's jazz and marching bands. A true American, he liked guns. Recently he'd purchased a Ruger forty-caliber pistol, and spent afternoons at Baron's Den gun range honing his target-shooting skills.

Whenever I saw him at Allen, he was usually moving with purpose to some class or interview, video camera in hand. For a guy six foot three, 220 pounds, he had a skeeter-bug energy to him. He talked fast, walked fast, entered and left the room fast. He was unfailingly upbeat. "What's the point in being negative? It's a waste of time," he told me. "Instead of worrying, you could use the energy *doing* something. Being proactive." At the moment, Prapat was proactively seeking a PR job with the Detroit Redwings. Hockey was another of his interests.

Aaron, Josh, Prapat. Korean, Filipino, Thai-Chinese. All sons of immigrants. All looking ahead with varying degrees of eagerness. All fully invested in the American way.

And yet. There was an "and yet" I could not overlook. If these young men were better off than earlier generations, they were likely not as well off as future generations will be. The in-between space had to do with belonging and yet not *fully* belonging. Perhaps it came with being part of a still-small minority. Five percent did not approach anything like a critical mass. There were still too many white and black Americans who looked upon Asians as exotic, *from the outside.* As if they had a scarlet remnant of "other" still attached to them. Whatever the reason, each of these young men, when prodded, expressed a lingering sense of not having entered some important inner circle. A sense that, yes, they were allowed into the house but not into its most hallowed chambers. They were accepted as American, but perhaps not *quite* as American as others. Not yet fully one of "us."

They all had stories. Aaron shared a couple.

First story. Freshman year. Late-night bull session in the dorm lounge. Students drinking and laughing. Most were white. Talk turned to sex and who was dating whom, and who *wanted* to date whom. Subject turned to the most desirable men on campus. Who was *hot?* Women in the group began naming young men — most of them white, a few black.

"What about Asian guys. Any hot Asian guys?" one girl asked.

"There *is* no such thing as a hot Asian guy," another girl quipped.

Laughter. The girl who made the remark then noticed Aaron. She feigned shock. Awkward silence in the room. "Except you," she said. A lame attempt to cover herself. Aaron smiled, but he got the message.

Most Asian American men continue to hear variations of the message. Prospects have improved, yes. Asian males are no longer the categorical rejects they once were; but as a whole, they still don't have the sexual cachet that other groups of men have. Josh told me that he felt no romantic interest from white women, who made up by far the largest segment of the female population on campus. His last girlfriend was Native American. Prapat said his dating life was practically "nonexistent." He'd tried online dating, but his queries drew no serious responses.

Second story. Junior year. Football practice. Aaron was a wide receiver who often ended up in the hinterlands of the field. After one play, he was the last to run up to what he thought was an informal huddle. One of the coaches had a formation in mind and told Aaron that he was standing in the wrong place. "Behind the line," the coach yelled. Which line? Aaron had not heard the first part of the instructions and did not know what the coach was talking about. The coach got in his face. "Behind the line," he barked. "In America, we play *American* football. Got it?"

Some of his teammates consoled Aaron after practice, including the team's star running back, Kenjon Barner, who told him, "Don't pay attention to him." Aaron didn't let the coach's words sink in until later. Aaron mulled them. He, who was born in Tacoma and raised in the suburbs among strip malls and McDonald's, who spent his four high school years slam-dunking as a Decatur Golden Gator, pledged allegiance to the flag, prayed to Jesus every Sunday, and afterward watched the Seahawks on television. He was as American as anybody. Wasn't he?

In the spring of 2011, a blond UCLA student named Alexandra Wallace posted a three-minute rant on YouTube about the "hordes of Asians" at her school. She was highly annoyed by them. While at the library, "I'll be typing away furiously," she says, "and then all of the sudden, when I'm about to, like, reach an epiphany, over here from somewhere, 'Ohh, Ching chong ling long ting tong? *Ohhhh.*'" Within hours, Wallace's video went viral, drawing more than a million views. One *LA Weekly* blogger called her an "overnight celebrity." Her critics eventually hounded her off the UCLA campus. I've overheard similar rants for years, most often spoken in private. Just this year, as I was sitting in a dining hall at UO, a white student at the next table complained about "the invasion" of Asians on campus. Only outsiders invade.

Aaron had encountered these sentiments, too. Yet, despite the handful of incidents that gave him pause, he did not seem discouraged in any enduring way. He made friends with the young woman from the late-night rap session at the dorm. He forgave the coach. He grinned at being part of the invasion. He appeared much more able to cope than I at his age, and to let negative experiences recede into the back corners of his mind — sometimes even explaining why some white and black Americans say what they do: "They just aren't used to seeing Asians. That's all it is."

The difference between us may be upbringing or basic disposition. He was a glass-half-full kind of guy. But I could not help thinking that at least part of Aaron's (and Josh's and Prapat's) emotional resiliency had to do with growing up in a very different America. True, there were still pockets of suspicion and hostility. And Asians were still a relatively small minority, and Asian women were still objectified and Asian men belittled. But it was undeniable that today's America was overall a more hospitable place.

In the Puget Sound region, where Aaron was born and raised, Asian American doctors, lawyers, teachers, ministers, police officers, and business owners were much more plentiful than when my family lived there. Entire neighborhoods from Tacoma to Everett were composed of Asian immigrants and their first- or second- or third-generation Asian American relatives. Aaron was one year old when Filipino American Velma Veloria became the first woman elected to the Washington State legislature. He was five when Chinese American Gary Locke became governor. He was nine when Japanese phenom Ichiro Suzuki reached rock-star status.

Aaron could go online or turn on the television and see that Asians were making their mark on the national scene, too. He was in elementary school when Chinese American Michelle Kwan established herself as one of the best figure skaters of all time; when stand-up comedian Margaret Cho, a Korean American, took her sardonic humor to stages across the country; when

Taiwanese American Jerry Yang of Yahoo! was recognized as one of the country's youngest and richest innovators. A man who was half Thai, Tiger Woods, rose to become the best golfer on the planet. Woods claimed to be also part Dutch, African American, and American Indian, which prompted him to refer to his ethnic make-up as "Cablinasian," an abbreviation for Caucasian, black, Indian, and Asian.

Aaron was ten when Chinese American Elaine Chao became U.S. secretary of labor. That same year, Chinese hoopster Yao Ming blazed into the NBA, dwarfing little guys like Shaquille O'Neal and Dirk Nowitzki. He was fourteen when Korean American actor Daniel Dae Kim, of the television show *Lost*, was named one of the sexiest men alive by *People* magazine, and he was fifteen when another Korean American, Yul Kwon, won the competitive reality television show *Survivor: Cook Islands.* The show had controversially divided teams into racial tribes, pitting blacks, whites, Asians, and Hispanics against one another. The winner would need brains as well as brawn. When the tall, muscular Kwon, who also happened to be a Yale-educated lawyer, emerged as the winner, he told an interviewer that his driving goal was to dispel the stereotype of Asians as "geeky." One commentator made up a new term in Kwon's honor: *smartthrob.*

Aaron was sixteen when Gary Locke joined two other Asian American men — Steven Chu, a Nobel Prize-winning physicist, and Eric Shinseki, a four-star army general — in Barack Obama's cabinet. Aaron was eighteen when the ubiquitous local boy Locke left his cabinet post to become the first Chinese American ambassador to China. And Aaron was nineteen, already slam-dunking for his high school team, when Jeremy Lin became a sensation with the New York Knicks and then signed a three-year, $25 million deal with the Houston Rockets. A home-grown Asian guy paid $25 million to play basketball!

I have to think that the bud of Aaron's potential found nurture in the accomplishments of individuals who resembled himself, found succor in the widening presence of the East in his Western surroundings.

At UO, Aaron could mingle with the hundreds of international students from China, Korea, Japan, India, and Vietnam. Many of them gathered at The Break pool hall on the lower level of the student union building during weekday afternoons. He could stop at any number of campus newsstands and pick up a copy of *HuaFeng,* a magazine written in both English and Mandarin. He could step into the Jordan Schnitzer Museum and look at architectural fragments from Cambodia's Khmer Empire or a luminous standing Buddha from Thailand. He could walk into Gerlinger Hall, headquarters of the new Confucius Institute for Global China Studies, and apply to an exchange program with the institute's co-sponsor, East China Normal University in Shanghai. He could take a short bus ride to the Lane County Fairgrounds and check out the annual Asian Celebration festival, where he could learn tai chi or watch an exhibit on making sushi or take part in *tambulilingan,* a traditional dance from Bali depicting love between a honeybee and a flower.

Asia penetrated deeply, even in a state as overwhelmingly white (89 percent) as Oregon. I had to wonder whether I would have felt like such an exile if I had grown up under these circumstances.

A little while ago, wandering in a Macy's across town, I found myself facing a mural-sized poster for Levi's 508 Taper Fit jeans. The square-jawed Asian guy modeling the 508s stood with arms casually at his sides, legs slightly spread apart, feet firmly planted, eyes looking unflinchingly at *me*. I've since happened upon Square Jaw in three different department stores in three different states. On a recent afternoon, coming from one of these stores, I attended a parent-teacher meeting at my daughter's middle school. I learned that the only ethnic-related student clubs were the origami and Korean clubs. There was no Irish or Italian or German club. No clubs for Saudi Arabians, Mexicans, or Somalis. The Korean Club, I was told, had a thriving membership and was highly active.

I could turn on cable television and watch twenty-four-hour programming on the Filipino Channel. There were dedicated channels for Japanese, Chinese, Koreans, Vietnamese, and South Asians. My cable service offered an on-demand film collection called "Cinema Asian America." One featured film, *Mr. Cao Goes to Washington*, told the true story of Vietnamese immigrant Joseph Cao, a lawyer and Republican who surprised the establishment in 2009 by getting elected congressman in a predominantly black Democratic district in New Orleans. It was a long way from Hop Sing and Bruce Lee.

Six months had passed since I watched Aaron play basketball at the rec center. I ran into him outside Agate Alley Bistro. The temperature was in the nineties, and he was wearing shorts and a tank top, and I found myself confronted by the slab of his body. I did not remember his shoulders being so wide and his arms and legs so muscular. He'd been working out with the football team every day, drinking protein shakes and lifting weights. Summer training, he said, smiling. A respectful, modest smile.

We went inside and visited for a while.

He ordered a plate of chicken tenders and French fries, which he devoured, it seemed to me, in about twenty seconds. He covered his mouth when he was chewing and speaking at the same time. We talked about his summer classes. We talked about how there was nothing to do in Eugene in July. His only plan for the weekend was to try a new church. In a couple of weeks, he was going to drive up to Federal Way to visit his girlfriend, a pretty young Korean American he had known for a long time. "It's the only thing to look forward to," he said.

Maybe it was the easy camaraderie, and the fact that his gentleness and openness had grown on me over the past year. At that moment, it just seemed as if his face was perfect. Innocent and strong. And—why not?—exotic, too, but exotic in the sense of being distinct from most of the other faces we saw every day. A face I would design if I could design a face for the son I was never going to have. It was an Asian face. Not unlike the one I used to see in the mirror so many decades ago, the one I was so ashamed of for its distinctive contours. The one I spent years trying to alter with clothespins and duct tape.

I was tempted to tell him about my clothespin adventures, but thought otherwise. Why drag the kid down into my ancient pit? He might not have been able to relate to the hole which I've imagined swallowing so many generations and which I've spent so much of my life trying to rise out of. His children and grandchildren may never know such a hole ever existed. Or if they learn about it, it will be as history. I will be an artifact.

14

Big Little Fighter

We are our memory,
we are that chimerical museum of shifting shapes,
that pile of broken mirrors.

— *Jorge Luis Borges*

Recently I got the chance to travel back to Cebu, to the island of Mactan, where Magellan met his end and where this story began. A couple of decades had passed since my first trip. The world had flipped its orbit in the intervening years, so rapidly had changes come. Long-held ways of seeing had become obsolete, and I had many more moments of wondering whether I'd wasted too much time pursuing phantoms.

Race. Manhood. What did these even mean in the twenty-first century? Words such as "manhood" and "manliness" were uttered with irony, as if the ideas had become obsolete. And definitions of "race" had splintered in so many directions, each leading to its own dimly lit maze. Two people using the same small word in a discussion today may not have any idea, may not even be able to grasp, what the other is talking about. I've over-heard many such discussions; I've engaged in more than a few. Often, they don't resemble discussions so much as evasions, or contests of noise.

In recent years, the one direction I've gone regularly was west, crossing the Pacific until I landed in the East. I found every excuse. My ulterior motive always was to grasp another piece of the puzzle that I had been working on for so long, as much out of habit now as anything else. But Mactan was an outback, a place you revisit only when life drops you in the vicinity. A journalism fellowship dropped me in Manila, forty-five flying minutes from Cebu. I went at the first chance, landing on a day not unlike the one so many years earlier, hazy and sticky-hot with the far horizon blurring into mirage.

Cebu City, still a human swarm, had more of the feel of burst-ing at the seams than I remembered. The winding roads seemed more crowded with storefronts and vendors and "standbys," the colloquial term for a person, usually male, who stands around doing nothing. If you asked, they might say, *Walang trabaho!* No job! The standbys had multiplied. The child beggars roaming in packs had grown more desperate. The squatter settlements had built up, cardboard shacks on top of scrap-tin huts, held to-gether by twine, a few scavenged nails, and the obstinate hopes of the people living there. The settlements had grown into cities, fragile as dried mud. When storms swept through, the settle-ments simply washed away, as happened on neighboring islands with Typhoon Haiyan in November 2013.

This was the other Asia. The one that the West heard less and less about as the Asian economic miracle played out in China and a handful of other "tiger economies." The one that tens of millions of people still lived and died in. The forlorn one that, after just an hour of looking around from the back seat of a taxi, made me turn around and lock myself in my hotel room with a bottle of Tanduay rum. That's how I spent my first day.

On the second day, I focused on faces. Faces were simple. Looking at individual faces narrowed my vision, kept me from getting lost in the dispiriting sprawl. They gave me refuge from my own thoughts, told me what I needed to know. At some point I began noticing the billboards and store displays and handbills featuring faces that resembled those of the people who saw them every day. Golden brown faces, Asian faces. That the lookers and the looked upon matched was, to me, something new.

I recalled, on my first trip, gazing up at the giant images of Richard Gere and Julia Roberts smiling from their billboard perches. Theirs and other white faces dominated the public square, as if the country were still a Western outpost, which in many ways it was.

Now, in traveling those same streets, the face I noticed most was Manny's. In this country, his last name was superfluous; only one Manny mattered. Manny Pacquiao was the most famous man in the Philippines and possibly the most famous Filipino who ever lived. In sporting circles he was regarded as not just the greatest prizefighter to come out of Asia but pound for pound one of the best prizefighters on the planet. His name was whispered among the names of the all-time greats—Henry Armstrong, Sugar Ray Robinson, Muhammad Ali. Relish did not begin to describe the sentiment of Filipinos on the subject. It amounted to a fervent savoring that was almost religious.

In so many of the ragged parts of the globe, manhood still found its most accessible expression in the persona of the warrior. To underdogs everywhere, to people in the developing world who are reminded every day of the smallness of their lives, Manny's appeal was his ability to render size meaningless. During his ten-year ascent to the top of the sport, he beat bigger men as easily as he beat smaller ones. He repeatedly proved the old maxim true about the bigger ones falling harder. Manny had started as a flyweight, moved through an astonishing eight weight divisions, and wrecked all comers. In December 2008 he defied the odds and pummeled the celebrated American boxer Oscar De La Hoya into submission and permanent retirement.

An on-air exchange by the stunned HBO announcing team:

"Pacquiao is the most exciting little fighter in the world."

"*Little!?* He looks big tonight!"

"*Big* little fighter in the world."

On the day of his matches, all of the Philippines stopped: government ground to a halt, business shut down, crime dropped to zero, combatants in the insurgency lay down their arms. Bandits and soldiers sat around the same hand-crank radio and for a few moments rooted for the same cause. And even though he lost twice in 2012 and showed the signs of a fading fighter, Manny was still seen by Filipinos as their Liu Xiang, their Yao Ming, their Ichiro. He was the gate-crasher, the buster of barriers, the countryman who went west and conquered. He was living evidence of the potential that languished in this tropical backwater with so many underfed bodies.

It had to make you wonder what would happen if all those bodies were fed. How many Manny Pacquiaos lived in the squatter villages or ran with the packs of spindly street kids on Archbishop Reyes Avenue? He was once one of them, weaving barefooted through traffic and selling cigarettes for a few centavos apiece. Now his triumphant likeness gazed upon the same streets, swarmed by a different generation of kids. Perhaps one or two would fight their way off the avenue just on the force of hope from a single glimpse of Manny's face on a bare concrete wall.

On Mango Avenue I searched out the restaurant where I'd encountered Mr. Shar-pei and his teenage escort so many years earlier. It was a different establishment now, hardly recognizable. Still, I did not have to search long on Mango to see the familiar couplings of older white men with their Filipinas. I no longer sneered at the sight. My indignation had softened into sadness, and curiosity had become more my style. One morning, I got to talking to one of these couples.

Peter and Mar-Len were staying in the same hotel. One morning, as I read the newspaper in the lobby, they sat near me, waiting for a ride. I guessed Peter was in his fifties. He was round in the middle, losing his hair. He wore a Hawaiian shirt and khaki shorts, and his skin had the burnt-pink hue common to foreigners here. His face was friendly. I'd learn that he was born in Berlin. Mar-Len looked to be in her early thirties, with bright eyes and skin the shade of hazelnut. She had a motherly way, even with strangers. Within minutes of meeting me, she exhorted me to "eat something" when I told her I hadn't had breakfast.

Peter was trying to recall where they were going that day. "It's toward the northern part of Cebu, I believe. Over by . . . where is it again, honey?"

"It's not *toward* the north, it *is* the north. The tip," Mar-Len said.

"Oh, of course. That's right. It's quite far, and the name is difficult. What was the name again, honey?"

"Daanbantayan," Mar-Len said. *"Dah-Ahn-Bahn-Tye-Ahn."*
Then to me: "He's been here four times and he still can't pronounce anything." Then to him, "Isn't that right, my German?"

"I have a ways to go," he said sheepishly.

"My German," she said. "What am I going to do with my German."

He reached over and clasped her hand. "You're going to take care of me, honey."

Over several conversations and e-mails with Mar-Len, I learned she was a single mother raising two kids. The father of her children had left to find work in Davao and was never heard from again. She met Peter through "a friend of a friend" who had put them in touch on the Internet. I suspected some kind of matchmaking service. When I met them they had known each other going on five years. Peter claimed to be divorced and living alone in a suburb of Hamburg. He sent Mar-Len regular "help" in the form of remittances. He visited the islands once a year and stayed for as long as a month.

"I love him. He's a good man," Mar-Len told me.

My younger self would have found their relationship objectionable. And there were aspects of their arrangement that, when I thought about them too long, still made me feel uneasy. Peter had so many more choices, and so much more power. He was a Westerner and therefore wealthy by local standards. He was royalty to Mar-Len's peasant class. For all Mar-Len knew, Peter could be married and living with kids of his own in Germany, could be sending remittances to other women on other islands, even in other countries. He could be visiting them just after he left Mar-Len. And there was no way for her to find out. He could fly all over the world, and she could not even leave her island. She did not have the means. And she may not have cared what Peter did once he left. It benefited her not to care.

I'd become more forgiving of the vagaries of relationships. People date, mate, marry, and join together in all kinds of ways for all kinds of reasons. If Mar-Len gave Peter comfort and company, and he gave her money and devotion, even for just one month out of the year, and neither objected nor came away damaged, who was I to judge them? If Mar-Len did not protest, how could I do so with any hope of being heard? When last I checked in, Peter was having a small concrete house built to re-place Mar-Len's old wooden one, which had been damaged in a monsoon. It cost him less than a thousand dollars. Mar-Len would have the most modern house in her *barangay*. She was grateful beyond words.

In one of our early conversations, she told me that she had dreamed of meeting an American, but a German was almost as good. "Just a foreigner," she said. "Of course, *diba?* We all want foreigners." "Foreigner" was code for "white" in these islands. She said she had given up on Filipino men; they had never come through for her. She thought foreigners "so handsome" and "so strong." She was convinced they were smarter, too. And kinder.

Kinder? I thought.

I was tempted to challenge her, but held back. At one time her words would have been salt rubbed into an old wound. But new skin had grown over the wound, and I could take Mar-Len's words without feeling as if my own worth as a man was at stake. Part of what constituted this new protective membrane was an idea introduced to me by a black writer named Frantz Fanon, who had died two years after I was born. Stumbling upon his work at the library was one of those quietly pivotal moments that I would look back upon with awe, the way one does with fateful meetings. A single chapter of one of his books knocked me out, and he's been knocking me out ever since and healing me in the process.

Fanon was a descendant of African slaves, born and raised in the French colony of Martinique, one of seven islands in the Caribbean known as the French Antilles. The islands were tropical, like those of the Philippines: beaches of various colors of sand, palm trees and lush tropical flora circled by lapping blue ocean. Most Antilleans are descended from African slaves brought there to work the sugar plantations.

In his 1952 book *Black Skin, White Masks*, Fanon devotes a chapter to the widespread phenomenon of black women obsessing over white men in France's Caribbean and African colonies.

He tells of one Antilles woman who had decided that she could love only a white man: *She asks for nothing, demands nothing, except for a little whiteness in her life. And when she asks herself whether he is handsome or ugly, she writes: "All I know is that he had blue eyes, blond hair, a pale complexion, and I loved him.* Another passage in the chapter says that black women were *obsessed with the dream of being wedded to a white man from Europe. . . . They must have a white man, a proper white man, and nothing but a white man. Almost all of them spend their entire lives waiting for this stroke of luck, which is anything but likely.*

These women also knew what they did not desire. *"I don't like the black man because he's a savage,"* says one woman. *"Not a savage in the cannibal sense but because he lacks refinement."* Another woman is more direct: *"I hate niggers. Niggers stink. They're dirty and lazy. Don't ever mention niggers to me."* And a young, educated black woman says, "*There is a white potential in every one of us; some want to ignore it or simply reverse it. Me, I would never accept to marry a nigger for anything in the world.*"

If the rejection of their own men happened (and continues to happen) among black women in the former colonies, and among yellow and brown women in those parts of the globe recently dominated by Westerners, then perhaps the rejection was less about color than about its benefits. To have "a little whiteness" represented the chance to experience power and privilege, and whoever came to possess these became more beautiful. The color (or ethnicity or race) of the rejected men represented hardship and vulnerability, which made the men *undesirable*. That word again! The one that had bedeviled me as a young man. We rejects were dismissed through no fault of our own, aside from having been born into a conquered tribe.

This happened time and again: I would explore a new path in my investigation of worthy manhood, and find that black men had already traveled it. In the New World, Africans in great numbers preceded Asians by three and a half centuries. Black men have struggled longer and harder against the forces that would keep them pinned to the bottom of the hole.

"I will be a man and know myself to be one, even among those who secretly and openly deny my manhood," wrote W.E.B. Du Bois, the first American black to earn a doctorate from Harvard.

So I have come to look to the African American story as a field guide through the swampier territories. The latest chapter of that story has jolted me awake to the possibilities. When Barack Obama was elected president in 2008, which I had thought impossible in 2007, I found my imagination roaming where it hadn't dared go before. Might there come a viable Asian American candidate for president within my lifetime? My children's? It would require a huge leap in cultural consciousness, but Americans have shown the capacity for such leaps.

Perhaps by then, though, the whole concept of race may be turned upside down (again) or further chopped up into disparate fragments. We already struggle to keep track of what we mean when we talk about race. Does the word refer to clusters or clines? Biological categories or cultural classifications? Political entities or medical classifications? Sociological groupings or regional populations? Are we talking about skin color, hair type, nose and eye shape, or skeletal formations? Nationality, ethnicity, political affiliation, or geographic origin? Distinct groups or groupable variations on a continuum?

None of the above or some of the above some of the time? Will one apply in one situation, another in a different situation? Or perhaps we will, by virtue of the word's kaleidoscopic nature, continue to change what we mean depending upon who we are talking to, or how we are twisting the scope on that particular day.

Western intelligentsia spent centuries explaining the reality of races — physically discrete human populations with given characteristics and in different stages of evolution, some more advanced than others. Then, in a long paroxysm of remorse, scholars and scientists spent the last half of the twentieth century claiming that race was a purely cultural construct with no biological basis. The matter seemed settled once and for all in 2000, when an international team completed a map of the human genome, and the head of the Human Genome Project, geneticist Francis Collins, declared, with President Bill Clinton by his side, that "the concept of race has no genetic or scientific basis."

Case closed. Race was dead!

Or was it? Many groups begged to differ. Clusters of geneticists, evolutionary biologists, forensic anthropologists, immunologists, epidemiologists, sports physiologists, and sociologists insisted that, among other things, the concept was useful in their line of work. That it had practical value.

Forensic anthropologists, for instance, contended that human beings seem generally to be grouped according to distinct sets of physical features that point to certain geographical origins, which help in identifying corpses. Those geographic origins seem to align with commonly accepted categories of race. Law enforcement insisted on using race in descriptions of suspects in an effort simply to narrow the universe, and they could show a long history of effectiveness. Epidemiologists offered compelling evidence that some diseases target certain population groups more than others, and in more than a few cases those populations roughly correspond with recognized racial groups — of vital importance to those needing or researching cures. And so on.

Just two years after the grand announcement, geneticist Collins changed his tune, writing in a medical journal, "It is not strictly true that race or ethnicity has no biological connection." Then, on a radio show, he said, "There are two points you can make about race and genetics. One is that we're really all very much alike. Incredibly alike. But you could also say even that small amount of difference turns out to be revealing."

That small amount of difference will be enough, it's my guess, for everyday people in everyday conversations to continue using racial categories for a long time to come — the rest of my life and probably well beyond. And it won't just be everyday people doing this. President Obama, responding to claims that racism fueled the mounting criticism against him, famously quipped to David Letterman, "It's important to realize that I was actually black before the election." And Clinton, even after his rousing speech on "our common humanity" at the completion of the genome project, has on several public occasions referred to himself as "a white guy."

For the time being, and until we collectively move on to more enlightened ways of identifying ourselves, I guess I am an Asian guy. I'm also numerous other things — writer, teacher, slacker; mystic, meditator, sinner; husband, father, prodigal; immigrant, citizen, dissenter; Filipino, American, Other — but I am at first contact, to countless people and institutions, an Asian. Even when I forget, and I forget more and more as my attention drifts and as life gives me other aspects of identity on which to focus, someone or something will remind me.

Recently I went to a new dentist and was asked to fill out the usual forms. And there it was, on page one, the space devoted to identifying my race. "Please check the box that most applies to you," it said. I checked the appropriate box, and only fleetingly wondered what it had to do with the crack I suspected in my upper-right third molar. Apparently my new dentist was part of the cohort that took exception to Francis Collins's human genome declaration. Whatever else I was thinking that day, whichever of my identities occupied me, there at the dentist's office, I was compelled to remember what racial group I belonged to. A very similar list of boxes appeared at my gastroenterologist's before my first colonoscopy, at the DMV when I applied for a new driver's license, etc. They present themselves at regular intervals.

I'm reminded even while in Asia. In Cebu, a food vendor near my hotel began speaking to me in the local dialect, Visayan, which I didn't understand. When I identified myself as

American, he said in heavily accented English, "Oh, sorry. You don't look like an American!" I asked him what I looked like. "You know. Asian. *Pilipino.*" Yep, I am. Later in the week, I asked a waiter about the fish in the "Fish Special" on the menu. He seemed reluctant to clarify.

"What kind of fish?"

"Just fish."

"Okay. Is it saltwater or freshwater?"

"Just . . . fish."

"But—"

That's when a Filipino companion turned and scolded me in a whisper. I didn't think I was asking anything unreasonable—I'm interested in where food comes from—but she thought I was putting on airs. "Stop acting like a foreigner. You're not a foreigner," she said. *"Ang iyong dugo ay mula dito.* Your blood is from here. Don't ever forget that."

In case you didn't notice, that was a woman scolding me at the restaurant, and a younger one at that. What she said was upsetting, but it didn't concern me that it was a woman saying it. Because, you see, a real man in the modern world accepted female authority without feeling diminished. What it meant to "be a man" had shifted dramatically, too, and continues to shift. America and the sections of the planet that have begun to give women their just due are being reshaped by the process. Being manly in these more progressive lands today meant recognizing women as equal shareholders in the human enterprise while also hanging on to one's balls.

Being strong and effective were still masculine ideals, but no longer exclusive to men, and no longer exclusively in the form of the bruising, no-holds-barred mid-twentieth-century way of

John Wayne. Just as race has splintered into a multiplicity of definitions, manhood has variegated to include an ever-widening spectrum of qualities. An effective man may exhibit some of the old Wayne machismo when push comes to shove, but his power could just as likely come from his pocketbook or position in the corporate hierarchy, from his wit and intellectual vigor or his creative passion. It could spring from his newly embraced vulnerability. Men could beat drums to the ancient primal rhythms, plumb the depths of grief, and weep in front of women and children. They could play the saxophone as evidence of soul. Just as Teddy Roosevelt and his trophy heads may have embodied an American ideal in his time, Bill Clinton saying "I feel your pain" and playing jazz sax embodied a model in *his* time.

Barak Obama, the most powerful man in the world and one of its most respected, told an audience that his most important role was not commander in chief but *father*, and that for him, life "revolves around my two little girls." Mature and benevolent fatherhood seems to require in men the combination of all the qualities I've described, both the hard and the soft, the exterior and the interior virtues. It requires the imposition of will when necessary, and the sublimation of ego when appropriate.

This makes me wonder whether the West, without knowing it, has slowly come around to the ancient Chinese view of *wen wu*. This view, you recall, held as ideal a man with martial ability but who in everyday dealings exhibited a softer masculinity tempered by study. This view venerated humility, mercy, and restraint. It held up wisdom and compassion as the highest attainments. The West has transposed its culture onto the East through the force of arms and the greater force of its popular culture. Might the traditional views of the East have infiltrated the West through the slower, less discernible process of cultural osmosis? Is it possible that the multilayered views of manhood forming in the West are being shaped in part by invisible undercurrents from the East?

Magellan was finally generating cash flow. All along the Mactan shoreline were new upscale resorts with shuttle service to the Magellan monument. The monument grounds, now called the Mactan Shrine, had been developed into a park with a formal entrance, concrete paths, gift shops, food stalls, historical markers, and gardens of native plants and trees identified by species — the kinds of things tourists look for.

In the few hours I was at the shrine, tour buses came and went nonstop. Most were filled with Chinese tourists, not something you would have seen a few decades ago. Now it seemed almost everyone at the park was Chinese. Not Chinese Filipinos, who are quite numerous in the Philippines, but Chinese from China, speaking in Chinese, carrying tourist guides written in Chinese, and in many cases even bringing their own food in Chinese-made portable containers.

Filipinos have wrung their hands for years over the growing Chinese presence in their country. Some have referred to the islands as a Chinese colony in the making. Ethnic Chinese dominate the national economy. One scholar calculates that pure ethnic Chinese (born of two Chinese parents) make up only 1 percent of the Philippine population but control 60 percent of the economy. Of the country's fifteen billionaires in 2012, nine were ethnic Chinese, six of them born in China. Ethnic Chinese are seen as intensely clannish, with Chinese merchants trading, networking, and strategizing almost exclusively with their *guanxi*, or traditional connections. Chinese enrich other Chinese, while the indigenous population languishes.

The same pattern has emerged in countries all over Southeast Asia. If the West sees China as a coming juggernaut, China's neighbors see it as a massive merchant army that has already invaded and won.

It was the sentiment that day among Filipino workers at the Mactan Shrine. A few expressed an open disdain for *Intsik*, the Filipino word for Chinese. "*Putang Ina ng Intsik!* They're rich now so they act rich," one worker told me, glancing around to make sure no one else heard him. "They tell you, 'Come here,' or 'Get that,' like they're your boss. They're not my fucking boss. Get out of here! *Intsik!*"

I watched Chinese tourists stroll the grounds while shielded from the sun by giant umbrellas held by Filipino boys trotting alongside them. Alabaster has long been the preferred skin color of Chinese elites, and now of elite wannabes. Their light complexions contrasted with the mocha hues of the Filipino workers. The umbrella boys worked hard that day, trotting to the pace of their masters, wiping their sweat with rags tucked in their pants, a wan solicitude in their eyes. A single good tip could buy enough rice for a week.

One group of Chinese tourists asked me to take their picture. They had six cameras among them and wanted a picture taken with each. "One more, one more," they kept saying, as they changed positions, laughing and having fun. Afterwards, a handsome couple in the group asked if I could throw away their trash, which they had been carrying in plastic bags. They didn't wait for my response before handing me the bags. Then it struck me: they thought I worked there.

Filipinos and Indonesians were the favored servants in much of rapidly developing China. Wealthy Chinese recruited from Philippine and Indonesian cities to find live-in housecleaners, nannies, gardeners, and chauffeurs. I once spent an afternoon in Hong Kong meeting with some of the thousands of Filipino domestic workers who gathered for fellowship every Sunday in the city's Statue Square. Something like 150,000 of them lived in Hong Kong. They made up an ethnic underclass. They sat around on benches and blankets, eating sack lunches and sharing stories, their only opportunity all week to laugh and speak in their own language.

Was the ancient hierarchy reemerging? The Chinese used to think of themselves as citizens of the Celestial Empire, to which all other peoples must pay tribute. Foreign peoples were elevated by their deference to the imperial throne. The Ming dynasty in the time of Zheng He turned away from the rest of the world because it believed it had nothing to learn from outsiders. I recalled an economics professor in Fujian who spoke to me of China's inevitable return to the top of the world order. It was just a matter of time.

"I have teapots older than your country," the professor said, sipping his tea. "And they're *very* sturdy teapots."

I've spent a lot of space in these pages aligning myself with people from other Asian countries and describing our common subjugation and generally nasty treatment at the hands of European and American colonizers. Now I'm reminded that Asians since ancient times have held to their own caste system — hierarchies that put their own kingdom on top of all others. It's possible that some kingdoms with which I've partnered myself in this quest for worthy manhood may want no association with me.

I descend from no great civilization. I come from one of Asia's outbacks. My people make up part of the continent's servant class, "Asia's niggers," as I've heard us described. I had drawn inspiration from the story of the great Chinese admiral Zheng He, but perhaps Zheng He's descendants might glance at me and see the outlines of a houseboy.

The Magellan spire looked exactly the same, still covered in a thin film of soot and surrounded by iron fencing, although the iron looked recently painted. Clusters of tourists stood around taking photographs. It was an odd-looking monument, like a fragment of a whole, as if the steeple of an old stone church had pierced the ground from beneath, and now the lone slender point stood in the middle of a cultivated garden. I walked around the spire, trying different vantage points. There was something concisely beautiful about it. Its simple, straight lines gave it elegance. The wear of time on the native stone gave it character. Even the thin layer of soot was like the dignified gray of an elder.

The Spaniards had built it in 1886 knowing they wouldn't be rulers of these islands forever, that they would eventually leave it in the hands of the native people. Sure enough, the Philippine fight for independence began ten years later. Soon after, the Spanish colonials fled, but Filipinos allowed the spire to remain standing, not just unvandalized but tended, protected. Now it stood as the centerpiece of a hallowed site. That may have been the source of its beauty — the story of an invasion, its bloody ending now sanctified, made part of the native ground, the invaders and defenders now granules on the same beach. Magellan was part of the Filipino story, the blood of his people now mixed with island blood, coursing through the same streams over centuries. Magellan, you could say, was now Filipino. He was part of my story, too. He coursed through my veins.

During graduate school I had a roommate named José, a Spanish national from Barcelona. He was about ten years my senior, an engineering student with pale skin and thick square glasses before they became fashionable. He was an introvert and a fellow brooder. Once, he told me that he did not consider Magellan an admirable figure, and that he was ashamed of the legacy of Spanish colonialism. "We Spaniards fucked the world," he told me. "We literally went everywhere, fucked everybody." I had not thought about José in years, and hadn't recalled those conversations in a couple of decades.

The recollection came as part of a deepening realization. Most of us inherit one shame or another. I had started to connect the far-flung dots of my many acquaintances. I met a guy named Mike during my travels in Alaska. We've stayed in touch. He's blond and blue-eyed, and does not fit comfortably in most chairs and beds because he's six foot nine. He's often embarrassed by his height and sometimes tells people he's six foot eight. He stoops on purpose. Once, while waiting to be seated at a restaurant, he and I stood in front of a full-length mirror. His reflection wasn't all there. His head was cut off. There were so many ways to be invisible.

I have another friend named Ken who has always been ashamed of his high IQ. It always set him apart. Being "too smart," he thought, was the main reason he'd been lonely most of his life. I've listened in on conversations in which he purposely dumbed himself down. I know too many women ashamed of being too fat, too thin, too short, too plain. I have men and women friends who've told me in all seriousness that they were ashamed of being white. They felt guilty, undeserving. During my stay in Cebu, I met a visiting European who told me she was ashamed of just one part of her whiteness. She was half English, half German, but disavowed her German side because, "well, do I need to explain why?" For a long time the notion that someone could be ashamed of being part of a conquering tribe was a hard one to wrap my mind around, but I get it now. We all have holes to climb out of.

Fifty paces north of the Magellan spire, closer to the shallows where most of the historic battle took place, stood a bronze statue of Lapu Lapu, the warrior chieftain who led the defense of the island. In some versions of the story, it was Lapu Lapu himself who thrust the spear into Magellan's face, finally killing him. A nearby plaque described him as "the first Filipino to have repelled European aggression."

The bronze figure, sixty-six feet tall, stood in a pose of martial vigilance, a sword in one hand, a shield in the other, the body heavily muscled, the face looking out toward the water. The face was the most memorable part for me. No Europeanized features there, no hint of a blue-eyed Jesus lurking in the shadows. It was a very Cebuano face, with distinctly Asian eyes, a broad nose, full lips, jutting cheekbones, square jaw and chin. It was mostly guesswork, of course, derived from local legend. But the face was highly plausible. It was an indigenous face, born of these islands. A face I could relate to.

Maybe it's all I would have needed as a boy looking for hope in myself: to look long at a hallowed face that could be my own, that invited me to a forgotten sense of home. It's a tempting thought, simple and romantic. I want to think now that my period of looking up from the bottom of a hole, and striving to climb out, was necessary to my journey somehow. And, rising above it, I would come to see that the nature of the hole itself was not as I imagined, that climbing out was a matter not simply of entering into manhood but of becoming the *particular* man — the genetically distinct, poetically unique, historically exact self—that I was fated to be. Perhaps seeing the image of Lapu Lapu, who seemed to grasp who and why he was, might have brought me to this point sooner. Might have.

Recently I ran across a paragraph describing how the artistic vision of French painter Georges Braque was transformed by seeing a single painting by Pablo Picasso, *Les Demoiselles d'Avignon.* The contemporary poet David Whyte tells of watching a Jacques Cousteau documentary as a boy and thereafter spending much of his young manhood pursuing the dream of being a marine biologist. I recall an interview with Sugar Ray Leonard in which he described seeing, as a youngster, a single round of one of Muhammad Ali's fights and thereafter knowing what kind of fighter he was going to be. A relative of mine watched the movie *The Mission* and left his comfortable life in America to be a Christian missionary among squatters in the rural Philippines. And I just told you how a single chapter in a book by Frantz Fanon enlarged my understanding of a particular phenomenon and in a very real way set me free. In each of these examples, timing was crucial: exposure had to occur when the seeker was able to receive the revelation.

If a painting or a chapter or a snippet of video had that power, could a single face have given me what I needed during all those glum hours of seeking a worthy origin and a worthwhile destiny? I imagine now that it could have made a difference to me as a young boy in Los Angeles and Seattle, or a teenager in the Bronx, or a young man in Oregon and Alaska seeking a place to belong and permission to stretch out. To become a human of worth. I mull the notion that something so simple could have launched me sooner, and a few feet farther, into a wide-open life.

CODA

My Family's Slave

"My Family's Slave" was written three years after Big Little Man *and was published in the* Atlantic *shortly after Alex Tizon's death, in March 2017.*

The ashes filled a black plastic box about the size of a toaster. It weighed three and a half pounds. I put it in a canvas tote bag and packed it in my suitcase this past July for the transpacific flight to Manila. From there I would travel by car to a rural village. When I arrived, I would hand over all that was left of the woman who had spent fifty-six years as a slave in my family's household.

Her name was Eudocia Tomas Pulido. We called her Lola. She was four foot eleven, with mocha-brown skin and almond eyes that I can still see looking into mine — my first memory.

She was eighteen years old when my grandfather gave her to my mother as a gift, and when my family moved to the United States, we brought her with us. No other word but "slave" encompassed the life she lived. Her days began before everyone else woke and ended after we went to bed. She prepared three meals a day, cleaned the house, waited on my parents, and took care of my four siblings and me. My parents never paid her, and they scolded her constantly. She wasn't kept in leg irons, but she might as well have been. So many nights, on my way to the bathroom, I'd spot her sleeping in a corner, slumped against a mound of laundry, her fingers clutching a garment she was in the middle of folding.

To our American neighbors, we were model immigrants, a poster family. They told us so. My father had a law degree, my mother was on her way to becoming a doctor, and my siblings and I got good grades and always said "please" and "thank you." We never talked about Lola. Our secret went to the core of who we were and, at least for us kids, who we wanted to be.

After my mother died of leukemia, in 1999, Lola came to live with me in a small town north of Seattle. I had a family, a career, a house in the suburbs — the American dream. And then I had a slave.

At baggage claim in Manila, I unzipped my suitcase to make sure Lola's ashes were still there. Outside, I inhaled the familiar smell: a thick blend of exhaust and waste, of ocean and sweet fruit and sweat.

Early the next morning I found a driver, an affable middle-aged man who went by the nickname "Doods," and we hit the road in his truck, weaving through traffic. The scene always stunned me. The sheer number of cars and motorcycles and jeepneys. The people weaving between them and moving on the sidewalks in great brown rivers. The street vendors in bare feet trotting alongside cars, hawking cigarettes and cough drops and sacks of boiled peanuts. The child beggars pressing their faces against the windows.

Doods and I were headed to the place where Lola's story began, up north in the central plains: Tarlac province. Rice country. The home of a cigar-chomping army lieutenant named Tomas Asuncion, my grandfather. The family stories paint Lieutenant Tom as a formidable man given to eccentricity and dark moods, who had lots of land but little money and kept mistresses in separate houses on his property. His wife died giving birth to their only child, my mother. She was raised by a series of *utusans*, or "people who take commands."

Slavery has a long history on the islands. Before the Spanish came, islanders enslaved other islanders, usually war captives, criminals, or debtors. Slaves came in different varieties, from warriors who could earn their freedom through valor to household servants who were regarded as property and could be bought and sold or traded. High-status slaves could own low-status slaves, and the low could own the lowliest. Some chose to enter servitude simply to survive: in exchange for their labor, they might be given food, shelter, and protection.

When the Spanish arrived, in the 1500s, they enslaved islanders and later brought African and Indian slaves. The Spanish Crown eventually began phasing out slavery at home and in its colonies, but parts of the Philippines were so far-flung that authorities couldn't keep a close eye. Traditions persisted under different guises, even after the United States took control of the islands in 1898. Today even the poor can have *utusans* or *katulongs* ("helpers") or *kasambahays* ("domestics"), as long as there are people even poorer. The pool is deep.

Lieutenant Tom had as many as three families of *utusans* living on his property. In the spring of 1943, with the islands under Japanese occupation, he brought home a girl from a village down the road. She was a cousin from a marginal side of the family, rice farmers. The lieutenant was shrewd—he saw that this girl was penniless, unschooled, and likely to be malleable. Her parents wanted her to marry a pig farmer twice her age, and she was desperately unhappy but had nowhere to go. Tom approached her with an offer: she could have food and shelter if she would commit to taking care of his daughter, who had just turned twelve.

Lola agreed, not grasping that the deal was for life.

"She is my gift to you," Lieutenant Tom told my mother.

"I don't want her," my mother said, knowing she had no choice.

Lieutenant Tom went off to fight the Japanese, leaving Mom behind with Lola in his creaky house in the provinces. Lola fed, groomed, and dressed my mother. When they walked to the market, Lola held an umbrella to shield her from the sun. At night, when Lola's other tasks were done—feeding the dogs, sweeping the floors, folding the laundry that she had washed by hand in the Camiling River—she sat at the edge of my mother's bed and fanned her to sleep.

One day during the war Lieutenant Tom came home and caught my mother in a lie—something to do with a boy she wasn't supposed to talk to. Tom, furious, ordered her to "stand at the table." Mom cowered with Lola in a corner. Then, in a quivering voice, she told her father that Lola would take her punishment. Lola looked at Mom pleadingly, then without a word walked to the dining table and held on to the edge. Tom raised the belt and delivered twelve lashes, punctuating each one with a word. *You. Do. Not. Lie. To. Me. You. Do. Not. Lie. To. Me.* Lola made no sound.

My mother, in recounting this story late in her life, delighted in the outrageousness of it, her tone seeming to say, *Can you believe I did that?* When I brought it up with Lola, she asked to hear Mom's version. She listened intently, eyes lowered, and afterward she looked at me with sadness and said simply, "Yes. It was like that."

Seven years later, in 1950, Mom married my father and moved to Manila, bringing Lola along. Lieutenant Tom had long been haunted by demons, and in 1951 he silenced them with a .32-caliber slug to his temple. Mom almost never talked about it. She had his temperament — moody, imperious, secretly fragile — and she took his lessons to heart, among them the proper way to be a provincial *matrona:* You must embrace your role as the giver of commands. You must keep those beneath you in their place at all times, for their own good and the good of the household. They might cry and complain, but their souls will thank you. They will love you for helping them be what God intended.

My brother Arthur was born in 1951. I came next, followed by three more siblings in rapid succession. My parents expected Lola to be as devoted to us kids as she was to them. While she looked after us, my parents went to school and earned advanced degrees, joining the ranks of so many others with fancy diplomas but no jobs. Then the big break: Dad was offered a job in foreign affairs as a commercial analyst. The salary would be meager, but the position was in America—a place he and Mom had grown up dreaming of, where everything they hoped for could come true.

Dad was allowed to bring his family and one domestic. Figuring they would both have to work, my parents needed Lola to care for the kids and the house. My mother informed Lola, and to her great irritation, Lola didn't immediately acquiesce. Years later Lola told me she was terrified. "It was too far," she said. "Maybe your mom and dad won't let me go home."

In the end, what convinced Lola was my father's promise that things would be different in America. He told her that as soon as he and Mom got on their feet, they'd give her an "allowance." Lola could send money to her parents, to all her relations in the village. Her parents lived in a hut with a dirt floor. Lola could build them a concrete house, could change their lives forever. *Imagine.*

We landed in Los Angeles on May 12, 1964, all our belongings in cardboard boxes tied with rope. Lola had been with my mother for twenty-one years by then. In many ways she was more of a parent to me than either my mother or my father. Hers was the first face I saw in the morning and the last one I saw at night. As a baby, I uttered Lola's name (which I first pronounced "Oh-ah") long before I learned to say "Mom" or "Dad." As a toddler, I refused to go to sleep unless Lola was holding me, or at least nearby.

I was four years old when we arrived in the United States, too young to question Lola's place in our family. But as my siblings and I grew up on this other shore, we came to see the world differently. The leap across the ocean brought about a leap in consciousness that Mom and Dad couldn't, or wouldn't, make.

• • •

Lola never got that allowance. She asked my parents about it in
a roundabout way a couple of years into our life in America. Her
mother had fallen ill (with what I would later learn was dysen-
tery), and her family couldn't afford the medicine she needed.
"*Pwede ba?*" she said to my parents. *Is it possible?* Mom let out
a sigh. "How could you even ask?" Dad responded in Tagalog.
"You see how hard up we are. Don't you have any shame?"

My parents had borrowed money for the move to the United
States, and then borrowed more in order to stay. My father was
transferred from the consulate general in Los Angeles to the
Philippine consulate in Seattle. He was paid $5,600 a year. He
took a second job cleaning trailers, and a third as a debt collec-
tor. Mom got work as a technician in a couple of medical labs.
We barely saw them, and when we did they were often exhausted
and snappish.

Mom would come home and upbraid Lola for not cleaning
the house well enough or for forgetting to bring in the mail.
"Didn't I tell you I want the letters here when I come home?"
she would say in Tagalog, her voice venomous. "It's not hard *na-
man*! An idiot could remember." Then my father would arrive
and take his turn. When Dad raised his voice, everyone in the
house shrank. Sometimes my parents would team up until Lola
broke down crying, almost as though that was their goal.

It confused me. My parents were good to my siblings and
me, and we loved them. But they'd be affectionate to us kids one
moment and vile to Lola the next. I was eleven or twelve when I
began to see Lola's situation clearly. By then Arthur, eight years
my senior, had been seething for a long time. He was the one
who introduced the word "slave" into my understanding of what
Lola was. Before he said it, I'd thought of her as just an unfortunate member of the household. I hated when my parents yelled

at her, but it hadn't occurred to me that they — and the whole arrangement — could be immoral.

"Do you know anybody treated the way she's treated?" Arthur said. "Who lives the way she lives?" He summed up Lola's reality: Wasn't paid. Toiled every day. Was tongue-lashed for sitting too long or falling asleep too early. Was struck for talking back. Wore hand-me-downs. Ate scraps and leftovers by herself in the kitchen. Rarely left the house. Had no friends or hobbies outside the family. Had no private quarters. (Her designated place to sleep in each house we lived in was always whatever was left — a couch or storage area or corner in my sisters' bedroom. She often slept among piles of laundry.)

We couldn't identify a parallel anywhere except in slave characters on TV and in the movies. I remember watching a Western called *The Man Who Shot Liberty Valance.* John Wayne plays Tom Doniphon, a gunslinging rancher who barks orders at his servant, Pompey, whom he calls his "boy." *Pick him up, Pompey. Pompey, go find the doctor. Get on back to work, Pompey!* Docile and obedient, Pompey calls his master "Mistah Tom." They have a complex relationship. Tom forbids Pompey from attending school but opens the way for Pompey to drink in a whites-only saloon. Near the end, Pompey saves his master from a fire. It's clear Pompey both fears and loves Tom, and he mourns when Tom dies. All of this is peripheral to the main story of Tom's showdown with bad guy Liberty Valance, but I couldn't take my eyes off Pompey. I remember thinking, *Lola is Pompey, Pompey is Lola.*

One night when Dad found out that my sister Ling, who was then nine, had missed dinner, he barked at Lola for being lazy. "I tried to feed her," Lola said as Dad stood over her and glared. Her feeble defense only made him angrier, and he punched her just below the shoulder. Lola ran out of the room, and I could hear her wailing, an animal cry.

"Ling said she wasn't hungry," I said.

My parents turned to look at me. They seemed startled. I felt the twitching in my face that usually preceded tears, but I wouldn't cry this time. In Mom's eyes was a shadow of something I hadn't seen before. Jealousy?

"Are you defending your Lola?" Dad said. "Is that what you're doing?"

"Ling said she wasn't hungry," I said again, almost in a whisper.

I was thirteen. It was my first attempt to stick up for the woman who spent her days watching over me. The woman who used to hum Tagalog melodies as she rocked me to sleep, and when I got older would dress and feed me and walk me to school in the mornings and pick me up in the afternoons. Once, when I was sick for a long time and too weak to eat, she chewed my food for me and put the small pieces in my mouth to swallow. One summer when I had plaster casts on both legs (I had problem joints), she bathed me with a washcloth, brought medicine in the middle of the night, and helped me through months of rehabilitation. I was cranky through it all. She didn't complain or lose patience, ever.

To now hear her wailing made me crazy.

In the old country, my parents felt no need to hide their treatment of Lola. In America, they treated her worse but took pains to conceal it. When guests came over, my parents would either ignore her or, if questioned, lie and change the subject. For five years in North Seattle, we lived across the street from the Misslers, a rambunctious family of eight who introduced us to things like mustard, salmon fishing, and mowing the lawn. Football on TV. Yelling during football. Lola would come out to serve food and drinks during games, and my parents would smile and thank her before she quickly disappeared. "Who's that little lady you keep in the kitchen?" Big Jim, the Missler patriarch, once asked. A relative from back home, Dad said. Very shy.

Billy Missler, my best friend, didn't buy it. He spent enough time at our house, whole weekends sometimes, to catch glimpses of my family's secret. He once overheard my mother yelling in the kitchen, and when he barged in to investigate found Mom red-faced and glaring at Lola, who was quaking in a corner. I came in a few seconds later. The look on Billy's face was a mix of embarrassment and perplexity. *What was that?* I waved it off and told him to forget it.

I think Billy felt sorry for Lola. He'd rave about her cooking, and make her laugh like I'd never seen. During sleepovers, she'd make his favorite Filipino dish, beef *tapa* over white rice. Cooking was Lola's only eloquence. I could tell by what she served whether she was merely feeding us or saying she loved us.

When I once referred to Lola as a distant aunt, Billy reminded me that when we'd first met I'd said she was my grandmother.

"Well, she's kind of both," I said mysteriously.

"Why is she always working?"

"She likes to work," I said.

"Your dad and mom—why do they yell at her?"

"Her hearing isn't so good . . ."

Admitting the truth would have meant exposing us all. We spent our first decade in the country learning the ways of the new land and trying to fit in. Having a slave did not fit. Having a slave gave me grave doubts about what kind of people we were, what kind of place we came from. Whether we deserved to be accepted. I was ashamed of it all, including my complicity. Didn't I eat the food she cooked, and wear the clothes she washed and ironed and hung in the closet? But losing her would have been devastating.

There was another reason for secrecy: Lola's travel papers had expired in 1969, five years after we arrived in the United States. She'd come on a special passport linked to my father's job. After a series of fallings-out with his superiors, Dad quit the consulate and declared his intent to stay in America. He arranged for permanent-resident status for his family, but Lola wasn't eligible. He was supposed to send her back.

Lola's mother, Fermina, died in 1973; her father, Hilario, in 1979. Both times she wanted desperately to go home. Both times my parents said "Sorry." No money, no time. The kids needed her. My parents also feared for themselves, they admitted to me later. If the authorities had found out about Lola, as they surely would have if she'd tried to leave, my parents could have gotten into trouble, possibly even been deported. They couldn't risk it. Lola's legal status became what Filipinos call *tago nang tago*, or TNT—"on the run." She stayed TNT for almost twenty years.

After each of her parents died, Lola was sullen and silent for months. She barely responded when my parents badgered her. But the badgering never let up. Lola kept her head down and did her work.

My father's resignation started a turbulent period. Money got tighter, and my parents turned on each other. They uprooted the family again and again—Seattle to Honolulu back to Seattle to the southeast Bronx and finally to the truck-stop town of Umatilla, Oregon, population 750. During all this moving around, Mom often worked twenty-four-hour shifts, first as a medical intern and then as a resident, and Dad would disappear for days, working odd jobs but also (we'd later learn) womanizing and who knows what else. Once, he came home and told us that he'd lost our new station wagon playing blackjack.

For days in a row Lola would be the only adult in the house. She got to know the details of our lives in a way that my parents never had the mental space for. We brought friends home, and she'd listen to us talk about school and girls and boys and whatever else was on our minds. Just from conversations she overheard, she could list the first name of every girl I had a crush on from sixth grade through high school.

When I was fifteen, Dad left the family for good. I didn't want to believe it at the time, but the fact was that he deserted us kids and abandoned Mom after twenty-five years of marriage. She wouldn't become a licensed physician for another year, and her specialty, internal medicine, wasn't particularly lucrative. Dad didn't pay child support, so money was always a struggle.

My mom kept herself together enough to go to work, but at night she'd crumble in self-pity and despair. Her main source of comfort during this time: Lola. As Mom snapped at her over small things, Lola attended to her even more — cooking Mom's favorite meals, cleaning her bedroom with extra care. I'd find the two of them late at night at the kitchen counter, griping and telling stories about Dad, sometimes laughing wickedly, other times working themselves into a fury over his transgressions. They barely noticed us kids flitting in and out.

One night I heard Mom weeping and ran into the living room to find her slumped in Lola's arms. Lola was talking softly to her, the way she used to with my siblings and me when we were young. I lingered, then went back to my room, scared for my mom and awed by Lola.

Doods was humming. I'd dozed for what felt like a minute and awoke to his happy melody. "Two hours more," he said. I checked the plastic box in the tote bag by my side—still there—and looked up to see open road. The MacArthur Highway. I glanced at the time. "Hey, you said 'two hours' two hours ago," I said. Doods just hummed.

His not knowing anything about the purpose of my journey was a relief. I had enough interior dialogue going on. I was no better than my parents. I could have done more to free Lola. To make her life better. Why didn't I? I could have turned in my parents, I suppose. It would have blown up my family in an instant. Instead, my siblings and I kept everything to ourselves, and rather than blowing up in an instant, my family broke apart slowly.

Doods and I passed through beautiful country. Not travel-brochure beautiful but real and alive and, compared with the city, elegantly spare. Mountains ran parallel to the highway on each side, the Zambales Mountains to the west, the Sierra Madre Range to the east. From ridge to ridge, west to east, I could see every shade of green all the way to almost black.

Doods pointed to a shadowy outline in the distance. Mount Pinatubo. I'd come here in 1991 to report on the aftermath of its eruption, the second largest of the twentieth century. Volcanic mudflows called *lahars* continued for more than a decade, burying ancient villages, filling in rivers and valleys, and wiping out entire ecosystems. The *lahars* reached deep into the foothills of Tarlac province, where Lola's parents had spent their whole lives, and where she and my mother had once lived together. So much of our family record had been lost in wars and floods, and now parts were buried under twenty feet of mud.

Life here is routinely visited by cataclysm. Killer typhoons that strike several times a year. Bandit insurgencies that never end. Somnolent mountains that one day decide to wake up. The Philippines isn't like China or Brazil, whose mass might absorb the trauma. This is a nation of scattered rocks in the sea. When disaster hits, the place goes under for a while. Then it resurfaces and life proceeds, and you can behold a scene like the one Doods and I were driving through, and the simple fact that it's still there makes it beautiful.

A couple of years after my parents split, my mother remarried and demanded Lola's fealty to her new husband, a Croatian immigrant named Ivan, whom she had met through a friend. Ivan had never finished high school. He'd been married four times and was an inveterate gambler who enjoyed being supported by my mother and attended to by Lola.

Ivan brought out a side of Lola I'd never seen. His marriage to my mother was volatile from the start, and money—especially his use of her money—was the main issue. Once, during an argument in which Mom was crying and Ivan was yelling, Lola walked over and stood between them. She turned to Ivan and firmly said his name. He looked at Lola, blinked, and sat down.

My sister Inday and I were floored. Ivan weighed about 250 pounds, and his baritone could shake the walls. Lola put him in his place with a single word. I saw this happen a few other times, but for the most part Lola served Ivan unquestioningly, just as Mom wanted her to. I had a hard time watching Lola vassalize herself to another person, especially someone like Ivan.

But what set the stage for my blowup with Mom was something more mundane.

She used to get angry whenever Lola felt ill. She didn't want to deal with the disruption and the expense, and would accuse Lola of faking or failing to take care of herself. Mom chose the second tack when, in the late 1970s, Lola's teeth started falling out. She'd been saying for months that her mouth hurt.

"That's what happens when you don't brush properly," Mom told her.

I said that Lola needed to see a dentist. She was in her fifties and had never been to one. I was attending college an hour away, and I brought it up again and again on my frequent trips home. A year went by, then two. Lola took aspirin every day for the pain, and her teeth looked like a crumbling Stonehenge. One night, after watching her chew bread on the side of her mouth that still had a few good molars, I lost it.

Mom and I argued into the night, each of us sobbing at different points. She said she was tired of working her fingers to the bone supporting everybody, and sick of her children always taking Lola's side, and why didn't we just take our goddamn Lola, she'd never wanted her in the first place, and she wished to God she hadn't given birth to an arrogant, sanctimonious phony like me.

I let her words sink in. Then I came back at her, saying she would know all about being a phony, her whole life was a masquerade, and if she stopped feeling sorry for herself for one minute she'd see that Lola could barely eat because her goddamn teeth were rotting out of her goddamn head, and couldn't she think of her just this once as a real person instead of a slave kept alive to serve her?

"A slave," Mom said, weighing the word. "A *slave*?"

The night ended when she declared that I would never understand her relationship with Lola. *Never.* Her voice was so guttural and pained that thinking of it even now, so many years later, feels like a punch to the stomach. It's a terrible thing to hate your own mother, and that night I did. The look in her eyes made clear that she felt the same way about me.

The fight only fed Mom's fear that Lola had stolen the kids from her, and she made Lola pay for it. Mom drove her harder. Tormented her by saying, "I hope you're happy now that your kids hate me." When we helped Lola with housework, Mom would fume. "You'd better go to sleep now, Lola," she'd say sarcastically. "You've been working too hard. Your kids are worried about you." Later she'd take Lola into a bedroom for a talk, and Lola would walk out with puffy eyes.

Lola finally begged us to stop trying to help her.

Why do you stay? we asked.

"Who will cook?" she said, which I took to mean, *Who would do everything?* Who would take care of us? Of Mom? Another time she said, "Where will I go?" This struck me as closer to a real answer. Coming to America had been a mad dash, and before we caught a breath, a decade had gone by. We turned around, and a second decade was closing out. Lola's hair had turned gray. She'd heard that relatives back home who hadn't received the promised support were wondering what had happened to her. She was ashamed to return.

She had no contacts in America, and no facility for getting around. Phones puzzled her. Mechanical things—ATMs, intercoms, vending machines, anything with a keyboard—made her panic. Fast-talking people left her speechless, and her own broken English did the same to them. She couldn't make an appointment, arrange a trip, fill out a form, or order a meal without help.

I got Lola an ATM card linked to my bank account and taught her how to use it. She succeeded once, but the second time she got flustered, and she never tried again. She kept the card because she considered it a gift from me.

I also tried to teach her to drive. She dismissed the idea with a wave of her hand, but I picked her up and carried her to the car and planted her in the driver's seat, both of us laughing. I spent twenty minutes going over the controls and gauges. Her eyes went from mirthful to terrified. When I turned on the ignition and the dashboard lit up, she was out of the car and in the house before I could say another word. I tried a couple more times.

I thought driving could change her life. She could go places. And if things ever got unbearable with Mom, she could drive away forever.

Four lanes became two, pavement turned to gravel. Tricycle drivers wove between cars and water buffalo pulling loads of bamboo. An occasional dog or goat sprinted across the road in front of our truck, almost grazing the bumper. Doods never eased up. Whatever didn't make it across would be stew today instead of tomorrow — the rule of the road in the provinces.

I took out a map and traced the route to the village of Mayantoc, our destination. Out the window, in the distance, tiny figures folded at the waist like so many bent nails. People harvesting rice, the same way they had for thousands of years. We were getting close.

I tapped the cheap plastic box and regretted not buying a real urn, made of porcelain or rosewood. What would Lola's people think? Not that many were left. Only one sibling remained in the area, Gregoria, ninety-eight years old, and I was told her memory was failing. Relatives said that whenever she heard Lola's name, she'd burst out crying and then quickly forget why.

I'd been in touch with one of Lola's nieces. She had the day planned: when I arrived, a low-key memorial, then a prayer, followed by the lowering of the ashes into a plot at the Mayantoc Eternal Bliss Memorial Park. It had been five years since Lola died, but I hadn't yet said the final goodbye that I knew was about to happen. All day I had been feeling intense grief and resisting the urge to let it out, not wanting to wail in front of Doods. More than the shame I felt for the way my family had treated Lola, more than my anxiety about how her relatives in Mayantoc would treat me, I felt the terrible heaviness of losing her, as if she had died only the day before.

Doods veered northwest on the Romulo Highway, then took a sharp left at Camiling, the town Mom and Lieutenant Tom came from. Two lanes became one, then gravel turned to dirt. The path ran along the Camiling River, clusters of bamboo houses off to the side, green hills ahead. The homestretch.

I gave the eulogy at Mom's funeral, and everything I said was true. That she was brave and spirited. That she'd drawn some short straws, but had done the best she could. That she was radiant when she was happy. That she adored her children and gave us a real home — in Salem, Oregon — that through the '80s and '90s became the permanent base we'd never had before. That I wished we could thank her one more time. That we all loved her.

I didn't talk about Lola. Just as I had selectively blocked Lola out of my mind when I was with Mom during her last years. Loving my mother required that kind of mental surgery. It was the only way we could be mother and son — which I wanted, especially after her health started to decline, in the mid-'90s. Diabetes. Breast cancer. Acute myelogenous leukemia, a fast-growing cancer of the blood and bone marrow. She went from robust to frail seemingly overnight.

After the big fight, I mostly avoided going home, and at age twenty-three I moved to Seattle. When I did visit, I saw a change. Mom was still Mom, but not as relentlessly. She got Lola a fine set of dentures and let her have her own bedroom. She cooperated when my siblings and I set out to change Lola's TNT status. Ronald Reagan's landmark immigration bill of 1986 made millions of illegal immigrants eligible for amnesty. It was a long process, but Lola became a citizen in October 1998, four months after my mother was diagnosed with leukemia. Mom lived another year.

During that time, she and Ivan took trips to Lincoln City, on the Oregon coast, and sometimes brought Lola along. Lola loved the ocean. On the other side were the islands she dreamed of returning to. And Lola was never happier than when Mom relaxed around her. An afternoon at the coast or just fifteen minutes in the kitchen reminiscing about the old days in the province, and Lola would seem to forget years of torment.

I couldn't forget so easily. But I did come to see Mom in a different light. Before she died, she gave me her journals, which filled two steamer trunks. Leafing through them as she slept a few feet away, I glimpsed slices of her life that I'd refused to see for years. She'd gone to medical school when not many women did. She'd come to America and fought for respect as both a woman and an immigrant physician. She'd worked for two decades at Fairview Training Center, in Salem, a state institution for the developmentally disabled. The irony: she tended to underdogs most of her professional life. They worshipped her. Female colleagues became close friends. They did silly, girly things together—shoe shopping, throwing dress-up parties at one another's homes, exchanging gag gifts like penis-shaped soaps and calendars of half-naked men, all while laughing hysterically. Looking through their party pictures reminded me that Mom had a life and an identity apart from the family and Lola. Of course.

Mom wrote in great detail about each of her kids, and how she felt about us on a given day—proud or loving or resentful. And she devoted volumes to her husbands, trying to grasp them as complex characters in her story. We were all persons of conse-quence. Lola was incidental. When she was mentioned at all, she was a bit character in someone else's story. "Lola walked my be-loved Alex to his new school this morning. I hope he makes new friends quickly so he doesn't feel so sad about moving again . . ." There might be two more pages about me, and no other men-tion of Lola.

The day before Mom died, a Catholic priest came to the house to perform last rites. Lola sat next to my mother's bed, holding a cup with a straw, poised to raise it to Mom's mouth. She had become extra attentive to my mother, and extra kind. She could have taken advantage of Mom in her feebleness, even exacted revenge, but she did the opposite.

The priest asked Mom whether there was anything she wanted to forgive or be forgiven for. She scanned the room with heavy-lidded eyes, said nothing. Then, without looking at Lola, she reached over and placed an open hand on her head. She didn't say a word.

Lola was seventy-five when she came to stay with me. I was married with two young daughters, living in a cozy house on a wooded lot. From the second story, we could see Puget Sound. We gave Lola a bedroom and license to do whatever she wanted: sleep in, watch soaps, do nothing all day. She could relax—and be free—for the first time in her life. I should have known it wouldn't be that simple.

I'd forgotten about all the things Lola did that drove me a little crazy. She was always telling me to put on a sweater so I wouldn't catch a cold (I was in my forties). She groused incessantly about Dad and Ivan: my father was lazy, Ivan was a leech. I learned to tune her out. Harder to ignore was her fanatical thriftiness. She threw nothing out. And she used to go through the trash to make sure that the rest of us hadn't thrown out anything useful. She washed and reused paper towels again and again until they disintegrated in her hands. (No one else would go near them.) The kitchen became glutted with grocery bags, yogurt containers, and pickle jars, and parts of our house turned into storage for—there's no other word for it—garbage.

She cooked breakfast even though none of us ate more than a banana or a granola bar in the morning, usually while we were running out the door. She made our beds and did our laundry. She cleaned the house. I found myself saying to her, nicely at first, "Lola, you don't have to do that." "Lola, we'll do it ourselves." "Lola, that's the girls' job." Okay, she'd say, but keep right on doing it.

It irritated me to catch her eating meals standing in the kitchen, or see her tense up and start cleaning when I walked into the room. One day, after several months, I sat her down.

"I'm not Dad. You're not a slave here," I said, and went through a long list of slavelike things she'd been doing. When I realized she was startled, I took a deep breath and cupped her face, that elfin face now looking at me searchingly. I kissed her forehead.

"This is *your* house now," I said. "You're not here to serve us. You can relax, okay?"

"Okay," she said. And went back to cleaning.

She didn't know any other way to be. I realized I had to take my own advice and relax. If she wanted to make dinner, let her. Thank her and do the dishes. I had to remind myself constantly: *Let her be.*

One night I came home to find her sitting on the couch doing a word puzzle, her feet up, the TV on. Next to her, a cup of tea. She glanced at me, smiled sheepishly with those perfect white dentures, and went back to the puzzle. *Progress*, I thought.

She planted a garden in the backyard—roses and tulips and every kind of orchid—and spent whole afternoons tending it. She took walks around the neighborhood. At about eighty, her arthritis got bad and she began walking with a cane. In the kitchen she went from being a fry cook to a kind of artisanal chef who created only when the spirit moved her. She made lavish meals and grinned with pleasure as we devoured them.

Passing the door of Lola's bedroom, I'd often hear her listening to a cassette of Filipino folk songs. The same tape over and over. I knew she'd been sending almost all her money—my wife and I gave her $200 a week—to relatives back home. One afternoon, I found her sitting on the back deck gazing at a snapshot someone had sent of her village.

"You want to go home, Lola?"

She turned the photograph over and traced her finger across the inscription, then flipped it back and seemed to study a single detail.

"Yes," she said.

Just after her eighty-third birthday, I paid her airfare to go home. I'd follow a month later to bring her back to the United

States—if she wanted to return. The unspoken purpose of her trip was to see whether the place she had spent so many years longing for could still feel like home.

She found her answer.

"Everything was not the same," she told me as we walked around Mayantoc. The old farms were gone. Her house was gone. Her parents and most of her siblings were gone. Childhood friends, the ones still alive, were like strangers. It was nice to see them, but . . . everything was not the same. She'd still like to spend her last years here, she said, but she wasn't ready yet.

"You're ready to go back to your garden," I said.

"Yes. Let's go home."

Lola was as devoted to my daughters as she'd been to my siblings and me when we were young. After school, she'd listen to their stories and make them something to eat. And unlike my wife and me (especially me), Lola enjoyed every minute of every school event and performance. She couldn't get enough of them. She sat up front, kept the programs as mementos.

It was so easy to make Lola happy. We took her on family vacations, but she was as excited to go to the farmers' market down the hill. She became a wide-eyed kid on a field trip: "Look at those zucchinis!" The first thing she did every morning was open all the blinds in the house, and at each window she'd pause to look outside.

And she taught herself to read. It was remarkable. Over the years, she'd somehow learned to sound out letters. She did those puzzles where you find and circle words within a block of letters. Her room had stacks of word-puzzle booklets, thousands of words circled in pencil. Every day she watched the news and listened for words she recognized. She triangulated them with words in the newspaper and figured out the meanings. She came to read the paper every day, front to back. Dad used to say she was simple. I wondered what she could have been if, instead of working the rice fields at age eight, she had learned to read and write.

During the twelve years she lived in our house, I asked her questions about herself, trying to piece together her life story, a habit she found curious. To my inquiries she would often respond first with "Why?" Why did I want to know about her childhood? About how she met Lieutenant Tom?

I tried to get my sister Ling to ask Lola about her love life, thinking Lola would be more comfortable with her. Ling cackled, which was her way of saying I was on my own. One day, while Lola and I were putting away groceries, I just blurted it out: "Lola, have you ever been romantic with anyone?" She smiled, and then she told me the story of the only time she'd come close. She was about fifteen, and there was a handsome boy named Pedro from a nearby farm. For several months they harvested rice together side by side. One time, she dropped her *bolo* — a cutting implement — and he quickly picked it up and handed it back to her. "I liked him," she said.

Silence.

"And?"

"Then he moved away," she said.

"And?"

"That's all."

"Lola, have you ever had sex?" I heard myself saying.

"No," she said.

She wasn't accustomed to being asked personal questions. "*Katulong lang ako,*" she'd say. *I'm only a servant.* She often gave one- or two-word answers, and teasing out even the simplest story was a game of twenty questions that could last days or weeks.

Some of what I learned: She was mad at Mom for being so cruel all those years, but she nevertheless missed her. Sometimes, when Lola was young, she'd felt so lonely that all she could do was cry. I knew there were years when she'd dreamed of being with a man. I saw it in the way she wrapped herself around one large pillow at night. But what she told me in her old age was that living with Mom's husbands made her think being alone wasn't so bad. She didn't miss those two at all. Maybe her life would have been better if she'd stayed in Mayantoc, gotten married, and had a family, like her siblings. But maybe it would have been worse. Two younger sisters, Francisca and Zepriana, got sick and died. A brother, Claudio, was killed. What's the point of wondering about it now? she asked. *Bahala na* was her guiding principle. *Come what may.* What came her way was another kind of family. In that family, she had eight children: Mom, my four siblings and me, and now my two daughters. The eight of us, she said, made her life worth living.

None of us were prepared for her to die so suddenly.

Her heart attack started in the kitchen while she was making dinner and I was running an errand. When I returned she was in the middle of it. A couple of hours later at the hospital, before I could grasp what was happening, she was gone—10:56 p.m. All the kids and grandkids noted, but were unsure how to take, that she died on November 7, the same day as Mom. Twelve years apart.

Lola made it to eighty-six. I can still see her on the gurney. I remember looking at the medics standing above this brown woman no bigger than a child and thinking that they had no idea of the life she had lived. She'd had none of the self-serving ambition that drives most of us, and her willingness to give up everything for the people around her won her our love and utter loyalty. She's become a hallowed figure in my extended family.

Going through her boxes in the attic took me months. I found recipes she had cut out of magazines in the 1970s for when she would someday learn to read. Photo albums with pictures of my mom. Awards my siblings and I had won from grade school on, most of which we had thrown away and she had "saved." I almost lost it one night when at the bottom of a box I found a stack of yellowed newspaper articles I'd written and long ago forgotten about. She couldn't read back then, but she'd kept them anyway.

Doods's truck pulled up to a small concrete house in the middle of a cluster of homes mostly made of bamboo and plank wood. Surrounding the pod of houses: rice fields, green and seemingly endless. Before I even got out of the truck, people started coming outside.

Doods reclined his seat to take a nap. I hung my tote bag on my shoulder, took a breath, and opened the door.

"This way," a soft voice said, and I was led up a short walkway to the concrete house. Following close behind was a line of about twenty people, young and old, but mostly old. Once we were all inside, they sat down on chairs and benches arranged along the walls, leaving the middle of the room empty except for me. I remained standing, waiting to meet my host. It was a small room, and dark. People glanced at me expectantly.

"Where is Lola?" A voice from another room. The next moment, a middle-aged woman in a housedress sauntered in with a smile. Ebia, Lola's niece. This was her house. She gave me a hug and said again, "Where is Lola?"

I slid the tote bag from my shoulder and handed it to her. She looked into my face, still smiling, gently grasped the bag, and walked over to a wooden bench and sat down. She reached inside and pulled out the box and looked at every side. "Where is Lola?" she said softly. People in these parts don't often get their loved ones cremated. I don't think she knew what to expect. She set the box on her lap and bent over so her forehead rested on top of it, and at first I thought she was laughing (out of joy) but I quickly realized she was crying. Her shoulders began to heave, and then she was wailing — a deep, mournful, animal howl, like I once heard coming from Lola.

I hadn't come sooner to deliver Lola's ashes in part because I wasn't sure anyone here cared that much about her. I hadn't expected this kind of grief. Before I could comfort Ebia, a woman walked in from the kitchen and wrapped her arms around her, and then she began wailing too. The next thing I knew, the room erupted with sound. The old people — one of them blind, several with no teeth — were all crying and not holding anything back. It lasted about ten minutes. I was so fascinated that I barely noticed the tears running down my own face. The sobs died down, and then it was quiet again.

Ebia sniffled and said it was time to eat. Everybody started filing into the kitchen, puffy-eyed but suddenly lighter and ready to tell stories. I glanced at the empty tote bag on the bench, and knew it was right to bring Lola back to the place where she'd been born.

Author's Note

The stories here are based on recollections, journals, letters, and documents that my family managed to retain through all our migrations. I've corroborated my recollections with relatives and friends when possible. Our memories aligned most of the time, but not always, and I'm reminded of Tobias Wolff's line about memory having its own story to tell.

It was an aid to memory that I've kept a journal since I was sixteen. But the mother lode of information was my mother's diaries. She doesn't get nearly the amount of print in these pages that she deserves for the exceptional life she led and the abundant love she gave her children. She kept a daily diary for fifty-two years until her death in 1999. She bequeathed her diaries to me. They fill two steamer trunks. Her entries record weather, meals, conversations, routines, happenings, tragedies, the state of her heart, every grievance she experienced, and everything she knew about her children, which was more than I expected. She kept track of income and expenses down to the penny. She wrote clinical descriptions of every illness in the family. She noted every time she and my father had sex, and rated her orgasms with a star system. She Scotch-taped addendums to her diaries. Some years required a second or third notebook.

Her entries often make references to letters and documents in various boxes and files she kept throughout the house. Home always doubled as a repository for family artifacts. As I think of it now, it was from her that I learned to write in journals and to keep files. In addition to her diaries and files and miscellaneous boxes of mementos, she amassed 188 photo albums, many of them two to three inches thick, with detailed captions on the back of each picture. The collection on its own is a remarkable record of our family.

Some of the chapters in this book required journalistic-style research and interviewing, but the book is more a series of reflections. The intent was to chronicle a mostly interior journey while staying true to exterior events, a tricky endeavor given the way most transformative experiences happen with no clear beginning and ending. For narrative simplicity, in the first chapter I merged two Mactan trips into one. I translated numerous conversations from Tagalog to English. In some chapters I changed names or omitted surnames to protect privacy. I make liberal use of other people's expertise, particularly that of writers and scholars whose work helped me to understand and explain historical events and scientific developments about which I can claim no authority.

The truest story I could tell in relation to the grand themes of this book was my own. I do identify patterns and occasionally make statements that sweep broadly, but I don't speak for anyone else. I testify mainly to my own experience, in the hope that my words may resonate with others still looking for words.

Acknowledgments

I am grateful beyond words to Terry McDermott for the many years of encouragement and tactical advice (and expensive Scotch). Without him, I would not have met Paul Bresnick and Deanne Urmy, and without them, this book would not have come to be. My deepest thanks to all three. To writers and scholars whose work contributed in immeasurable ways: David Wellman, Kam Louie, Ronald Takaki, Richard Bernstein, Sheridan Prasso, Louise Levathes, Richard Steckel, Michael Kimmel, John Dower, Sam Keen, and Ian McNeely; to friends and colleagues who read or contributed to portions of the manuscript: Steve Podry, Tim Davis, Linda Keene, Brian Lindstrom, Mitchell Fox, Valentina Petrova, Lisa Heyamoto, Deb Merskin, Stephanie Essin, Nora Quiason, Janette Bustos, Lynn Marshall, Tim

Gleason, Cecilia Balli, Donald Katz, and Carlin Romano; and to the J. Anthony Lukas Prize Project, the East West Center, the International Center for Journalism, the Philippine Center for Investigative Journalism, and the University of Oregon School of Journalism and Communication; and to Lola, whom I miss every day — I say *Salamat*.

Selected Sources

1. Killing Magellan

Agoncillo, Teodoro A. *A Short History of the Philippines*. New York: New American Library, 1969.

Bergreen, Laurence. *Over the Edge of the World: Magellan's Terrifying Circumnavigation of the Globe*. New York: William Morrow, 2003.

Cameron, Ian. *Magellan and the First Circumnavigation of the World*. New York: Saturday Review Press, 1973.

Jocano, F. Landa. *Filipino Prehistory: Rediscovering Precolonial Heritage*. Quezon City: PUNLAD Research House, 1998.

Lawrence, D. H. "Song of a Man Who Has Come Through." In *The Complete Poems of D. H. Lawrence*, 195. Ware, Hertfordshire: Wordsworth, 1994.

Pigafetta, Antonio. *Magellan's Voyage: A Narrative Account of the First Circumnavigation*. New Haven: Yale University Press, 1969.

Sweig, Stefan, *Conqueror of the Seas: The Story of Magellan*. New York: Viking Press, 1938.

Zaide, Gregorio F. *The Pageant of Philippine History: From Prehistory to the*

Eve of the British Invasion. Manila: Philippine Education Company, 1979.

2. Land of the Giants

Constantino, Renato. *Neocolonial Identity and Counter-Consciousness.* London: Merlin Press, 1978.

Karnow, Stanley. *In Our Image: America's Empire in the Philippines.* New York: Ballantine Books, 1989.

Li Wei. "For the Heck of It." In *21st Century Chinese Poetry,* no. 3, ed. and trans. Meifu Wang and Steven Townsend, 43. Washington, D.C.: Pathsharers, 2012.

Memmi, Albert. *The Colonizer and the Colonized.* New York: Orion Press, 1965.

Murphey, Rhoads. *A History of Asia.* New York: Pearson Longman, 2006.

Pagden, Anthony. *Peoples and Empires: A History of European Migration, Exploration, and Conquest, from Greece to the Present.* New York: Modern Library, 2001.

Warren, James Francis. *Pirates, Prostitutes and Pullers: Explorations in the Ethno- and Social History of Southeast Asia.* Crawley, Australia: University of Western Australia Press, 2008.

3. Orientals

Chang, Iris. *The Chinese in America.* New York: Viking, 2003.

Dower, John W. *War Without Mercy: Race and Power in the Pacific War.* New York: Pantheon, 1987.

Hemingway, Ernest. *Winner Take Nothing.* New York: Charles Scribner's Sons, 1934.

London, Jack. *The Strength of the Strong.* London: Macmillan, 1914.

Park, Robert Ezra. *Race and Culture: Essays in the Sociology of Contemporary Man.* New York: Free Press, 1950.

Pfaelzer, Jean. *Driven Out: The Forgotten War Against Chinese Americans.* New York: Random House, 2007.

Said, Edward W. *Orientalism.* New York: Vintage, 1978.

Takaki, Ronald. *Strangers from a Different Shore: A History of Asian Americans.* Boston: Little, Brown, 1989.

Tanaka, Stefan. *Japan's Orient: Rendering Pasts into History.* Berkeley: University of California Press, 1994.

Vidal, Gore. "The Day the American Empire Ran Out of Gas." *The Nation,* January 11, 1986.

——. "The Empire Lovers Strike Back." *The Nation,* March 22, 1986.

Wu, Frank H. *Yellow: Race in America Beyond Black and White.* New York: Basic Books, 2002.

Zia, Helen. *Asian American Dreams: The Emergence of an American People.* New York: Farrar, Straus and Giroux, 2000.

4. Seeking Hot Asian Babes

Bernstein, Richard. *The East, the West, and Sex: A History of Erotic Encounters.* New York: Knopf, 2009.

De Mente, Boye. *Bachelor's Japan.* Rutland, Vt.: Charles E. Tuttle, 1967.

Evans, Karin. *The Lost Daughters of China: Abandoned Girls, Their Journey to America, and the Search for a Missing Past.* New York: Tarcher/Penguin, 2000.

Komroff, Manuel, ed. *The Travels of Marco Polo (The Venetian).* New York: Liveright Publishing, 1926.

Lehman, Peter, ed. *Pornography: Film and Culture.* New Brunswick, N.J.: Rutgers University Press, 2006.

Prasso, Sheridan. *The Asian Mystique: Dragon Ladies, Geisha Girls, and Our Fantasies of the Orient.* New York: Public Affairs, 2005.

Talmadge, Eric. "American GIs Frequented 'Comfort Women'; U.S. Military Complicit." Associated Press, April 25, 2007.

Tizon, Alex. "Death of a Dreamer." *Seattle Times,* April 21, 1996.

——. "Rapists Bet on Victims' Silence—and Lose." *Seattle Times,* May 31, 2001.

Woan, Sunny. "White Sexual Imperialism: A Theory of Asian Feminist Jurisprudence." *Washington and Lee Journal of Civil Rights and Social Justice* 13 (2008): 274–301.

Yosano Akiko. "V." In *One Hundred More Poems from the Japanese,* trans. Kenneth Rexroth, 7. New York: New Directions, 1974.

5. Babes, Continued

Chu, Ying. "The New Trophy Wives: Asian Women." *Marie Claire,* August 5, 2009.

Fisman, Raymond, Sheena S. Iyengar, Emir Kamenica, and Itamar Simonson. "Racial Preferences in Dating." *Review of Economic Studies* 75 (2008): 117–32.

Hwang, S., R. Saenz, and B. E. Aguirre. "Structural and Assimilationist Explanations of Asian American Intermarriage." *Journal of Marriage and the Family* 59 (1997): 758–72.

Le, C. N. "Inter-racial Dating and Marriage." *Asian Nation: Asian American History, Demographics, and Issues,* 2010. www.asian-nation.org.

Nemoto, Kumiko. *Racing Romance: Love, Power, and Desire Among Asian American–White Couples.* Piscataway, N.J.: Rutgers University Press, 2009.

Phillips, Sam. "Baby I Can't Please You." *Martinis and Bikinis.* Virgin, 1994. Audio CD.

Rivers, Tony. "Oriental Girls: The Ultimate Accessory." *Gentlemen's Quarterly* (British ed.), October 1990.

United States Census Bureau. "Decennial Census Data on Marriage and Divorce." 2000 and 2010. www.census.gov.

6. Asian Boy

Dower, John W. *Embracing Defeat: Japan in the Wake of World War II.* New York: W. W. Norton, 2000.

Eng, David L. *Racial Castration: Managing Masculinity in Asian America.* Durham: Duke University Press, 2001.

Hamamoto, Darrell. *Monitored Peril: Asian Americans and the Politics of TV Representation.* Minneapolis: University of Minnesota Press, 1994.

Huang, Tom. "Tasteless or Tone-Deaf?" April 2, 2004. www.poynter.org.

Kerouac, Jack. *The Dharma Bums.* New York: Penguin, 1971.

Mura, David. *Where the Body Meets Memory: An Odyssey of Race, Sexuality, and Identity.* New York: Anchor, 1997.

Yang, Wesley. "Paper Tigers: What Happens to All the Asian-American Over-achievers When the Test-Taking Ends?" *New York,* May 8, 2011.

7. Tiny Men on the Big Screen

Biggers, Earl Derr. *The House Without a Key: A Charlie Chan Mystery.* Indianapolis: Bobbs-Merrill, 1925.

Eagan, Daniel. "Reel Culture: Hollywood Goes to China." May 10, 2012. www.smithsonian.com.

Kashiwabara, Amy. "Vanishing Son: The Appearance, Disappearance, and Assimilation of the Asian American Man in American Mainstream Media." University of California, Berkeley, Media Resources Center. 2010.

Lee, Robert G. *Orientals: Asian Americans in Popular Culture.* Philadelphia: Temple University Press, 1999.

Marchetti, Gina. *Romance and the "Yellow Peril": Race, Sex, and Discursive Strategy in Hollywood Fiction.* Berkeley: University of California Press, 1994.

Moy, Ed. "Does Hollywood 'White-Wash' the Casting of Asian Characters in Movies?" July 29, 2009. www.examiner.com.

Slek, Stephanie. "Is Hollywood 'White-Washing' Asian Roles?" January 13, 2012. www.cnn.com.

Vargas, Jose Antonio. "'The Slanted Screen' Rues the Absence of Asians." *Washington Post,* March 25, 2007.

Winfrey, Yayoi Lena. "Asians on White Screens: Is Charlie Chan Really Dead?" 2001. www.IMDiversity.com.

8. Its Color Was Its Size

Baldwin, James. *Just Above My Head.* New York: Dial Press, 1979.

Buss, David M. *The Evolution of Desire: Strategies of Human Mating.* Rev. ed. New York: Basic Books, 2003.

Fisher, Helen. *Anatomy of Desire: A Natural History of Mating, Marriage, and Why We Stray.* New York: Ballantine, 1992.

Friedman, David M. *A Mind of Its Own.* New York: Penguin, 2003.

Gould, James L., and Carol Grant Gould. *Sexual Selection: Mate Choice and Courtship in Nature.* New York: Scientific American Library, 1997.

Kerouac, Jack. *On the Road.* New York: Viking Penguin, 1955.

Keuls, Eva C. *The Reign of the Phallus.* Berkeley: University of California Press, 1993.

Paley, Maggie. *The Book of the Penis.* New York: Grove Press, 1999.

Poulson-Bryant, Scott. *Hung: A Meditation on the Measure of Black Men in America.* New York: Doubleday, 2005.

Sheets, Connor Adam. "Jeremy Lin Manhood Size Discussions Reveals Racist Subtext in Linsanity." *International Business Times,* February 22, 2012.

Torre, Pablo S. "Against All Odds: The Sudden and Spectacular Ascent of Jeremy Lin — from Couch to Clutch." *Sports Illustrated,* February 20, 2012.

Wang, Meifu. "Dirt Road." In *21st Century Chinese Poetry,* no. 4, ed. and trans. Meifu Wang, Michael T. Soper, and Steven Townsend, 60. Washington, D.C.: Pathsharers, 2012.

9. Getting Tall

Baten, Joerg, Debin Ma, et al. "Evolution of Living Standards and Human Capital in China in 18th–20th Centuries: Evidences from Real Wage, Age Heaping, and Anthropometrics." *Explorations in Economic History* 47 (2010): 347–59.

Belot, Michele, and Jan Fidrmuc. "Anthropometry of Love: Height and Gender Asymmetries in Interethnic Marriages." *Economics & Human Biology* 8, no. 3 (December 2010): 361–72.

Bilger, Burkhard. "The Height Gap: Why Europeans Are Getting Taller and Taller — and Americans Aren't." *New Yorker,* April 5, 2004.

Dallek, Robert. *Flawed Giant: Lyndon Johnson and His Times, 1961–1973.* New York: Oxford University Press, 1999.

Demick, Barbara. "A Small Problem Growing: Chronic Malnutrition Has Stunted a Generation of North Koreans." *Los Angeles Times,* February 12, 2004.

Devereux, Stephen. *Famine in the Twentieth Century.* Brighton: Institute of Development Studies, University of Sussex, 2000.

Dikötter, Frank. *Mao's Great Famine: The History of China's Most Devastating Catastrophe, 1958–62.* New York: Walker & Co., 2010.

Eveleth, Phyllis B., and James M. Tanner. *Worldwide Variation in Human Growth.* Cambridge: Cambridge University Press, 1991.

Kimmel, Michael. *Manhood in America: A Cultural History.* New York: Free Press, 1996.

O'Brien, Edna. "A Conversation with Edna O'Brien." By Phillip Roth. *New York Times,* November 18, 1984.

Prasso, Sheridan. *The Asian Mystique.* New York: Public Affairs, 2005.

Qian Zheng. *China's Ethnic Groups and Religions.* Sinopedia Series. Cengage Learning Asia, 2012. Kindle ed.

Rossabi, Morris. *Governing China's Multiethnic Frontiers.* Seattle: University of Washington Press, 2005.

Usher, Rod. "A Tall Story for Our Time: In the Rise and Rise of Modern Mankind, Scientists Are Discovering That Height Bears a Clear Relationship to Healthiness and Social Well-Being." *Time,* October 14, 1996.

10. *Wen Wu*

Barry, Rob, Madeline Farbman, Jon Keegan, and Palani Kumanan. "Murder in America." *Wall Street Journal,* August 11, 2013.

Chang, Kuei-Sheng. "The Maritime Scene in China at the Dawn of Great European Discoveries." *Journal of the American Oriental Society* 94, no. 3 (1974): 347–59.

Cooper, Alexia, and Erica Smith. "Homicide Trends in the United States, 1980–2008." U.S. Department of Justice, Bureau of Justice Statistics. November 16, 2011.

Emerson, Ralph Waldo. *Prose and Poetry.* New York: Norton, 2001.

Gordon, Stewart. *When Asia Was the World: Traveling Merchants, Scholars, Warriors, and Monks Who Created the "Riches of the East."* Cambridge, Mass.: Da Capo Press, 2008.

Ho, Vanessa. "Voyage of Hope, Legacy of Sorrow." *Seattle Post-Intelligencer,* April 23, 2001.

Kristof, Nicholas D. "1492: The Prequel." *New York Times Magazine,* June 6, 1999.

Levathes, Louise. *When China Ruled the Seas: The Treasure Fleet of the Dragon Throne, 1405–1433.* New York: Simon & Schuster, 1994.

Louie, Kam. *Theorising Chinese Masculinity: Society and Gender in China.* 2002. Reprint, Cambridge: Cambridge University Press, 2009.

Louie, Kam, and Morris Low, eds., *Asian Masculinities: The Meaning and Practice of Manhood in China and Japan*. London: Routledge, 2003.

Maddison, Angus. *Contours of the World Economy, 1–2030 AD: Essays in Macro-Economic History*. New York: Oxford University Press, 2007.

Montgomery, L. M. *Emily Climbs*. London: Starfire, 1983.

Temple, Robert. *The Genius of China: 3,000 Years of Science, Discovery, and Invention*. New York: Simon & Schuster, 1986.

Tizon, Alex. "The Rush to 'Gold Mountain.'" *Seattle Times*, April 16, 2000.

Viviano, Frank. "China's Great Armada." *National Geographic*, July 2005.

Watts, Sarah. *Rough Rider in the White House: Theodore Roosevelt and the Politics of Desire*. Chicago: University of Chicago Press, 2006.

Winchester, Simon. *The Man Who Loved China: The Fantastic Story of the Eccentric Scientist Who Unlocked the Mysteries of the Middle Kingdom*. New York: HarperCollins, 2008.

11. Yellow Tornado

Jaques, Martin. *When China Rules the World: The End of the Western World and the Birth of a New Global Order*. New York: Penguin, 2009.

Mahbubani, Kishore. *The New Asian Hemisphere: The Irresistible Shift of Global Power to the East*. New York: Public Affairs, 2008.

McDonald, Mark. "Too Much Olympic Fever in China?" *International Herald Tribune*, August 7, 2012.

Saionji Kintsune. "XL." In *One Hundred Poems from the Japanese*, trans. Kenneth Rexroth, 40. New York: New Directions, 1964.

Vidal, Gore. *Imperial America: Reflections on the United States of Amnesia*. New York: Nation Books, 2005.

Yardley, Jim. "Racial 'Handicaps' and a Great Sprint Forward." *New York Times*, September 8, 2004.

Zakaria, Fareed. *The Post-American World*. New York: W. W. Norton, 2008.

12. "What Men Are Supposed to Do"

Kakinomoto no Hitomaro. "XXII." In *One Hundred Poems from the Japanese*, trans. Kenneth Rexroth, 24. New York: New Directions, 1964.

Semple, Kirk. "As Asian-Americans' Numbers Grow, So Does Their Philanthropy." *New York Times,* January 8, 2013.

Tizon, Alex. "A Fighter on the Fringe." *Seattle Times,* December 20, 1998.

———. "An Iraq War All His Own." *Los Angeles Times,* February 5, 2007.

———. "The Story of a Drive-By Murder." *Seattle Times,* March 8, 1998.

13. "One of Us, Not One of Us"

Pew Research Center, Social and Demographic Trends. "The Rise of Asian Americans." June 19, 2012.

Qian, Zhenchao, and Daniel T. Lichter. "Changing Patterns of Interracial Marriage in a Multiracial Society." *Journal of Marriage and Family* 73, no. 5 (October 2011): 1065–84.

Siegel, Lee. "Rise of the Tiger Nation." *Wall Street Journal,* October 27, 2012.

Swarns, Rachel L. "For Asian-American Couples, a Tie That Binds." *New York Times,* March 30, 2012.

Tu Fu. "To Wei Pa, a Retired Scholar." In *One Hundred Poems from the Chinese,* trans. Kenneth Rexroth, 11. New York: New Directions, 1971.

Viera, Mark. "For Lin, Erasing a History of Being Overlooked." *New York Times,* February 12, 2012.

Wang, Oliver. "Living with Linsanity." *Los Angeles Review of Books,* March 6, 2012.

Wang, Wendy. "The Rise of Intermarriage." Pew Research Center, Social and Demographic Trends. February 16, 2012.

Young, Susan. "Win Boosts Asian Image: 'Survivor' Leader from San Mateo Proud to Alter TV Stereotype." *San Jose Mercury News,* December 19, 2006.

14. Big Little Fighter

Abumrad, Jad, and Robert Krulwich. "Race." December 15, 2008. www.Radiolab.org.

Borges, Jorge Luis. "Cambridge." In *Selected Poems,* ed. Alexander Coleman, 271. New York: Penguin, 1999.

Chua-Eoan, Howard, and Ishaan Tharoor. "The Great Hope: Why Manny

Pacquiao Is More Than Just the World's Best Boxer." *Time* (Asia ed.), November 16, 2009.

Collas-Monsod, Solita. "Ethnic Chinese Dominate PH Economy." *Philippine Daily Inquirer,* June 22, 2012.

Dikötter, Frank. *Discourse of Race in Modern China.* New ed. Stanford: Stanford University Press, 1994.

Fanon, Frantz. *Black Skin, White Masks.* Trans. Richard Philcox. New York: Grove Press, 2008.

Harper, Phillip Brian. *Are We Not Men? Masculine Anxiety and the Problem of African American Identity.* New York: Oxford University Press, 1996.

Kahn, Jonathan. "Race in a Bottle." *Scientific American,* July 15, 2007.

Lewis, David Levering. *W.E.B. Du Bois: A Biography.* New York: Henry Holt, 1994.

Ochoa, Francis T. J. "Greatest Fighter of Era." *Philippine Daily Inquirer,* November 16, 2009.

Olson, Steve. *Mapping Human History: Genes, Race, and Our Common Origins.* Boston: Mariner Books, 2006.

Sarich, Vincent, and Frank Miele. *Race: The Reality of Human Differences.* Oxford: Westview Press, 2004.

Further Reading

Bederman, Gail. *Manliness and Civilization: A Cultural History of Gender and Race in the United States, 1880–1917.* Chicago: University of Chicago Press, 1995.

Boorstin, Daniel J. *The Discoverers: A History of Man's Search to Know His World and Himself.* New York: Random House, 1983.

Chua, Amy. *Day of Empire: How Hyperpowers Rise to Global Dominance — and Why They Fall.* New York: Anchor Books, 2007.

Debenham, Frank. *Discovery and Exploration: An Atlas-History of Man's Wanderings.* Garden City, N.Y.: Doubleday, 1960.

Espiritu, Yen Le. *Asian American Women and Men: Labor, Laws, and Love.* 2nd ed. Lanham, Md.: Rowman & Littlefield, 2008.

Fanon, Frantz. *The Wretched of the Earth.* 1961. Reprint, New York: Grove Press, 2005.

Fong-Torres, Ben. *The Rice Room: Growing Up Chinese American: From Number Two Son to Rock 'n' Roll.* New York: Plume, 1995.

Gilmore, David. *Manhood in the Making: Cultural Concepts of Masculinity.* New Haven: Yale University Press, 1990.

Hall, Stephen. *Size Matters: How Height Affects the Health, Happiness, and Success of Boys — and the Men They Become.* Boston: Houghton Mifflin, 2003.

Hanson, Victor Davis. *Carnage and Culture: Landmark Battles in the Rise of Western Power.* New York: Doubleday, 2001.

Huntington, Samuel P. *The Clash of Civilizations and the Remaking of World Order.* New York: Touchstone, 1997.

Jahoda, Gustav. *Images of Savages: Ancient Roots of Modern Prejudice in Western Culture.* London: Routledge, 1999.

Karnow, Stanley. *Vietnam: A History.* Rev. ed. New York: Penguin, 1997.

Keen, Sam. *Fire in the Belly: On Being a Man.* New York: Bantam, 1991.

Kim, Daniel Y. *Writing Manhood in Black and Yellow.* Stanford: Stanford University Press, 2005.

Landes, David S. *The Wealth and Poverty of Nations: Why Some Are So Rich and Some So Poor.* New York: W. W. Norton, 1999.

Lee, Gus. *China Boy.* New York: Plume, 1994.

Liu, Eric. *The Accidental Asian: Notes of a Native Speaker.* New York: Random House, 1998.

Mansfield, Harvey C. *Manliness.* New Haven: Yale University Press, 2006.

Merskin, Debra L. *Media, Minorities, and Meaning: A Critical Introduction.* New York: Peter Lang, 2011.

Molnar, Stephen. *Human Variation: Races, Types, and Ethnic Groups.* 6th ed. Upper Saddle River, N.J.: Prentice-Hall, 2005.

Morning, Ann. *The Nature of Race: How Scientists Think and Teach About Human Difference.* Berkeley: University of California Press, 2011.

Mura, David. *Turning Japanese: Memoirs of a Sensei.* New York: Anchor, 1991.

Nisbett, Richard E. *The Geography of Thought: How Asians and Westerners Think Differently . . . and Why.* New York: Free Press, 2003.

Painter, Nell Irvin. *The History of White People.* New York: W. W. Norton, 2010.

Preston, Diana. *The Boxer Rebellion: The Dramatic Story of China's War on Foreigners That Shook the World in the Summer of 1900.* New York: Walker & Co., 1999.

Takaki, Ronald. *A Different Mirror: A History of Multicultural America.* New York: Back Bay Books, 1993.